1:3 000 000 MAP AND

THE MOUNTAINS OF CENTRAL ASIA

COMPILED BY

The Royal Geographical Society

AND

The Mount Everest Foundation

THE MOUNTAINS
OF
CENTRAL ASIA

1:3 000 000 MAP AND GAZETTEER

THE MOUNTAINS
OF
CENTRAL ASIA

COMPILED BY
The Royal Geographical Society
AND
The Mount Everest Foundation

MACMILLAN
LONDON

First published 1987 by
MACMILLAN LONDON LIMITED
4 Little Essex Street London WC2R 3LF
and Basingstoke

Associated companies in Auckland, Delhi,
Dublin, Gaborone, Hamburg, Harare, Hong
Kong, Johannesburg, Kuala Lumpur, Lagos,
Manzini, Melbourne, Mexico City, Nairobi,
New York, Singapore and Tokyo

British Library Cataloguing in Publication Data

Mountains of Central Asia.
1. Mountains —— Asia, Central   2. Asia,
Central —— Description and travel.
I. Royal Geographical Society   II. Mount
Everest Foundation
915.8'04   DS785

ISBN 0-333-39600-6

Typeset by Universe Typesetters Limited
Printed by Anchor Brendon Ltd, Tiptree

CONTENTS

INTRODUCTION

The Region

The area covered by *Mountains of Central Asia* is one of the most diverse in the world, both geographically and culturally. It contains the world's highest mountain – Everest, or Qomolangma – and the lowest part of China – the Turfan depression, or Turpan Pendi; its vegetation varies from arctic to desert and tropical rain-forest; and, on the Tibetan plateau – the largest and highest in the world – rise most of the main rivers of China, India and south-east Asia. The ancient civilisation of Tibet is still to be seen in the traditional way of life that exists side by side with modern developments and advances in technology.

Origins

The name 'Tibet' derives from many sources – the Chinese *Turfan,* Arabic *T'ubett,* and Mongolian and Thai *Thibet* – indicating the mixed cultural heritage of the area. Legend tells that the country once lay at the bottom of a vast ocean, created by the flooding of the Tsangpo river, or Yarlung Zangbo Jiang, which covered most of the area; what land remained was primeval forest. Five gigantic poisonous dragons arose from the sea and destroyed the forest, and at that same moment five rose-coloured clouds appeared and were turned into five fairies who slew the dragons. The five fairies became the five main peaks of the Himalaya; Everest is one of them. (The ancient myth is founded on fact. Fossil remains of the fish-like reptile *Ichthyosaur* have been found on the mountain Xixabangma in the south and dinosaur remains occur in eastern Tibet. Fossil snails too have been found in the Himalaya.) Tradition also has it that the Buddhas cut through the south-eastern Himalaya; the waters of the region flowed through the gorge they made, and so the Tsangpo river was born.

According to myth, the people of Tibet are descended from the union of a macaque monkey and an ogress. Six little monkeys were born, and the tribe later multiplied to 500. Because they ate wild grains and the fruits of trees, their tails became short, they learned to speak and, in due course, turned into human beings. More prosaically, neolithic remains have been found in

Tibet, and in north Bhutan genetic studies of people of Tibetan origin show that there are not only affinities with the Tibetans of the plateau and races such as the Khasi of Assam (which would be expected) but also with the Ainu of Japan, Eskimo and North American Indians.

Geography

It is now known that the Central Asian plateau and its mountains are still being formed by the northward movement of the tectonic plate on which India lies. Many millions of years ago the Central Asian plateau was the eastern part of an ocean – the Tethys Sea – of which the western part remains as the Mediterranean. The plateau itself is being lifted up as a result of the collision of India and Asia. This is also resulting in some shortening of the earth's crust in a north–south direction, and, because the process is still continuing, the Himalaya are rising slowly at about 1–2cm a year. It is as though Asia were being crumpled, crushed and compressed by India.

The plateau has a total surface area of 1200 sq.km, with an average height of 5000m. It is 2600km from east to west and 1300km from north to south, and its geographical features are many and varied.

Its southern boundary is formed by the Himalaya, which include the world's highest mountains. It is about 2400km long, extending from Nanga Parbat in the west to Namjagbarwa, or Namche Barwa, in the east. The Indus, which curves round the western end, and the Tsangpo, which curves round the east, rise within a few miles of each other near Kangrinboqê, or Mt Kailas, a sacred peak in south Tibet. The northern boundary of the plateau is the Kunlun range. At its western side this is contiguous with the Pamir and Karakoram, while on its northern side is Taklimakan Shamo, or the Takla Makan Desert, in the Tarim Basin. North of this desert lies the Tien Shan, a mountain range running eastwards into Central Asia; Ürümqi, or Urumchi, the capital of Xinjiang (Sinkiang Province), lies towards their eastern end. The eastern part of the Kunlun has on its northern flank Qaidam Pendi (the Tsaidam Basin) – a salt-encrusted marshy area. North again is the Altun Shan (the Altyn Tagh range). Towards the north-eastern extremity of the plateau lies the lake Qinghai Hu (Koko Nor). Xining, the capital of Qinghai (Tsinghai Province) lies close to the lake. In central Tibet most of the ranges extend in an east–west direction – Hoh Xil Shan, to the south of the Kunlun, and the Tanggula Shan forming the dividing line between Qinghai and Tibet. The Nyainqêntanglha Shan lies just north of Lhasa. Further west are the Nganglong Kangri and Gangdisê Shan. In east and south-east Tibet the

Hengduan Shan is the name given to the ranges that run north–south beyond the bend of the Tsangpo and Namjagbarwa Feng.

Kangrinboqê is one of the most sacred peaks of southern Tibet, regarded as holy by both Hindus and Bhuddists. It lies north of two lakes, Mapam Yumco (Manasarowar) and La'nga Co (Rakas), which separate it from another peak, Gurla Mandhata. It was in this area that in 1936 two Swiss geologists, Heim and Gansser, found the first evidence of rocks from the ocean floor. Three important rivers rise close by: the Indus (Sengyo Kabab – 'Out of the Lion's Mouth'); the Ganges (Macha Kabab – 'Out of the Peacock's Mouth'); the Tsangpo (Tachok Kabab – 'Out of the Elephant's Mouth'). Other major rivers that rise on the plateau are the Chang Jiang (Yangtse), Lancang Jiang (Mekong) and Nu Jiang (Salween).

Tropical rain-forests and arctic tundra lie within a few miles of each other, resulting in an extraordinary variety of vegetation and flowers. Many species transported form south-east Tibet have been grown in European gardens; of these, *Meconopsis betonicifolia* (syn. *M. baileyi*) – the blue poppy – is the most spectacular. Lakes cover much of Tibet and support a great deal of wildlife. Three of the largest freshwater lakes are the Nam Co, Yamzho Yumco and Siling Co.

Agriculture and resources

By about 100 BC agriculture was established. Cross-breeding a yak and a cow produced the *dzo*, an animal better suited to the climate and altitude of Tibet than a pure yak. A system of irrigation to suit natural conditions was also developed. Elementary astronomy evolved as a servant of agriculture, to assist in finding the optimum time for ploughing, sowing and reaping.

Much of the plateau contains extensive grasslands and in northern Tibet up to 8 million cattle, sheep or goats graze. The grain lands of southern Tibet and the Chang Jiang, Lancang Jiang and Nu Jiang river valleys are the main agricultural districts, producing highland barley, wheat, peas, corn and millet. Apples, peaches and bananas have been introduced in the last twenty years in some areas, and rice is grown in south-east Tibet.

Surveys conducted in recent years have revealed a wealth of ferrous and non-ferrous metals. In west Tibet gold-mining has been carried out for centuries; in AD 450 Herodotus mentions gold-digging ants – almost certainly a reference to gold dust found in the loose earth burrowed by marmots. The biggest potential copper-mine in Asia has been found on the plateau.

Hydro-electricity is being developed as an important source of power and small-scale hydro-electric plants are now being built in certain areas.

Geothermal power sources are found in many areas, particularly at Yangbajain to the north of Lhasa, and there are numerous hot-water springs.

Climate

The southern barrier of the Himalaya shields the plateau from monsoon rains. Because of the altitude, the weather is generally cold, but the dry air and intense radiation from the sun cause rapid daily temperature changes. In the highest zone of the inner plateau, the Chang Tang (4500 – 5500m), the average temperatures in the hottest months are between 4 to 10°C with a daytime maximum of 25°C and a minimum of below freezing. In winter the average temperature of the coldest month varies from −15 to −10°C with a minimum of −40°C. Lhasa, at 3600m, has an average temperature in July of 20°C and in January of −3°C. By contrast, in some valleys of south-east and east Tibet the climate is almost tropical. Rainfall is poor except in the summer, and in the winter blizzards are common. Windstorms, with abrupt changes in temperature, occur throughout the year. Many large earthquakes have been recorded here.

Population

The population of Tibet is in the region of 1.7 million, the majority of whom are Tibetan. Many of the towns grew up around Buddhist monasteries; Lhasa, with about 120,000 inhabitants, is the largest. Other main centres are Xigazê (Shigatse), in the Tsangpo valley, and Qamdo, on the river Lancang Jang in east Tibet. Because of climatic conditions, south and east Tibet have the largest populations, particularly in the main valley of the Tsangpo and its tributary valleys.

Religion

Lying between the civilisations of Europe, China and India, Tibet developed its own brand of Buddhism. The influence of Islam has always been strong to the north of the Kunlun and in the Pamir, while it is thought that many centuries ago small groups of Nestorian Christians may have lived in Central Asia. Buddhism was introduced to Tibet in the seventh century by the king Songtsen Gampo, fusing with, rather than simply replacing, the primitive local Bon religion. Its influence has been profound and all-embracing. Temporal and spiritual powers are invested in one person, the Dalai Lama, whose residence is the Potala – a vast Buddhist 'Vatican' in Lhasa. This, one of the world's finest buildings, was begun in 1642 by the fifth Dalai Lama, Ngawang Lobzang Gyatso.

Medicine

Traditional Tibetan medicine, which is still practised in Lhasa, Xigazê and elsewhere, is a mixture of herbal cures and Buddhist custom. Yontan Gonpo, an outstanding physician of the eighth century, is considered to be the system's founder. Four medical treatises or mantras were introduced during this period; in them, all illness is attributed to an imbalance of the three humours that govern the human body – bile, phlegm and wind. The principal Tibetan medical institution, now a ruin, which also administered the astronomical almanac, was built on Chakpori Hill in Lhasa, opposite the Potala. More recently, a medical college, Mendzekhang, was built on the west side of the city, and there is also now a modern hospital. The main High-Altitude Research Unit is at Xining, the capital of Qinghai Province; although not on the plateau itself, it has outlying facilities. The Unit was opened in 1984 and, in addition to carrying out research, serves as a hospital.

Exploration
Early travellers

The mapping and gathering of information about the Central Asian plateau has been a tedious, complex, risky and romantic task.

India was known to Herodotus; he made no mention of mountains in the north, but Indian cosmographers sang the praise of Mt Kailas as the home of the god Siva. Ptolemy gave a masterly picture of India but, as he compressed the Himalaya and Kunlun into one range, Tibet did not exist, and such was his authority that for nearly 1500 years most Europeans did not know that there was a Central Asian plateau. However, geographical learning flourished with the Arabs, who, through the merchant Suleiman, knew of Tibet in 880.

The Chinese were the most knowledgeable about Central Asia because trade, politics and geographical proximity brought them into contact with the people of this region long before Europeans. Trade in spices and other commodities was almost as important as that in silk. The silk routes along which the Chinese travelled were among the longest and oldest trade routes in the world, for silk was one of the most prized commodities of the Romans. Skirting the northern side of the Tibetan plateau, these routes ran from the Caspian Sea to Xi'an in China and were conduits not only for trade but also for religion, culture, ideas and disease. Having crossed the Russian Pamir, the main route branched at Kashi (Kashgar), an oasis city at the western end of the Taklimakan Shamo; one way skirted the Tien Shan on the north side, while another went round the southern side under the Kunlun. The main centres on each of these routes were towns situated on rivers, which often dried up or changed course before they were lost in the salt marshes of the Lop Nur. The third branch ran south through the Gez gorge towards the Pamir plateau, passing Kongur and Muztagata before it plunged into the maze of ranges and valleys on its way to the Indian sub-continent. Lined with the bones of pack animals, yaks, goats, sheep and horses, this route was once a dirt track but is now a metalled road – the Karakoram highway.

Chang Ch'ien traversed Xinjiang in 126 BC, visiting the Amudar'ya (Oxus) and the country west of Kashi; Hsuan-Tsang, who has been termed the Prince of Chinese Pilgrims, travelled along the Silk Road to India in

AD 630. In AD 747 a Chinese expeditionary force crossed the Pamirs from Kashi to country to the west that paid tribute to the Middle Kingdom and from the fourth to the seventh centuries Buddhist monks from India travelled these routes to China.

In 1275 Marco Polo passed through the Pamir in 'forty good days', becoming the first European to give any valuable information about Tibet, although he did not actually visit the country. A decade later, 1322–4, a Franciscan ascetic, Odorico of Pardenone, visited Canbalac – the Beijing of today – and on his return may have travelled through Tibet.

Missions to Tibet

Tibet had to be rediscovered by Europe in the sixteenth and seventeenth centuries. Originally drawn to Central Asia by reports of pockets of Nestorian Christians, the Jesuit mission in India became seriously interested in trans-Himalayan exploration in 1581. With breathless enthusiasm, Father Aquaviva wrote: 'We have discovered a new nation of heathens called Bottan (Tibet) which is beyond Lahore towards the River Indus – a nation very well inclined and given to pious works. They are white men and Mohammedans do not live among them, wherefore we hope that if two sacred Fathers are sent thither, a great harvest of other heathens may be reaped'. However, the first mission was withdrawn from India in 1583. Eventually, with endorsement from the Pope and King Philip II of Spain, Bento de Goes was selected to lead such a mission to Tibet. Disguised as an Indian, he reached Shache (Yarkand) and then continued to Turpan near Ürümqi (the present capital of Sinkiang), Hami and Suchan. In identifying Cathay with China, he resolved a geographical mystery and also established the perimeter of Tibet, although he found no Christian enclaves. He was followed in 1624–8 by Father Antonio de Andrade. Accompanied by Manuel Marques, a lay brother, Andrade reached Badrīnāth (a Hindu shrine) and then made his way to Māna. Leaving Marques he crossed the Māna Pass and entered south Tibet – probably the first European to do so – setting up a mission at Zarang (Tsaparang), then one of the great cities of Tibet.

In 1626 Fathers Cacella and Cabral crossed from India into Bhutan – the first Europeans to make this journey. After some delays they reached Xigazê in south Tibet, but after a short period Cabral returned to India via Nepal. Cacella tried to reach Zarang but died of exhaustion before he got there.

It was not until 1661 that Fathers d'Orville and Grüber set out from

Beijing to return to Europe overland. Using an ancient caravan route, they approached Tibet by Xining – and three days later reached Qinghai Hu (Koko Nor), the great 'Blue Sea'. Crossing the Burhan Budai Shan (east Kunlun) they travelled across the 4000m barren and inhospitable plain that is now Qinghai Province. Crossing the Tanggula Shan by a 4500m pass, they reached Lhasa on 8 October 1661 – the first Europeans to do so. They were immediately struck by the size, architectural beauty and solidarity of the Potala rising abruptly from the edge of the plains. Their stay in Lhasa was brief – but Grüber sketched a picture of the fifth Dalai Lama, one of Tibet's greatest rulers, and was the first European to describe the prayer wheel. Travelling south from Lhasa they crossed the Thung La into Nepal and may well have been the first Europeans to see Everest. After spending some time in Kathmandu, they reached India. D'Orville died soon after at the Mogor Mission at Agra; Grüber returned to Rome via Persia and Turkey.

To see if any Christian converts remained at Zarang, another Jesuit, Father Desideri, entered Tibet, reaching Lhasa in 1716, and while there witnessed the invasion of Tibet by Dzungurs from northern Xinjiang.

Political struggles

By the end of the eighteenth century, three empires were competing for the political vacuum that was Central Asia and Tibet: Russia, China and Britain. The British, fearing the Russians in particular, were the most active and, following a local quarrel with the Bhutanese, Warren Hastings, newly appointed Governor of Bengal, decided 'to send an English Gentleman into Tibet for the purpose of exploring the unknown region, of opening material communications of trade and of making a treaty of friendship'. He chose George Bogle, a twenty-eight-year-old Scot who entered Bhutan and then Tibet, probably by the Tremo La and Phāri Dzong. It is worth noting that in Bhutan Bogle planted ten potatoes, the far-sighted Hastings having thought that this new product might eventually be introduced to the country; years later, descendants of the potatoes appeared in the bazaar at Lhasa. Bogle met the Tasho, or Panchen Lama, second only in importance to the Dalai Lama; they became very friendly and it was made clear to Bogle that the Chinese Emperor had a strong influence in Lhasa. Bogle was, however, never able to visit Lhasa or to see the Dalai Lama and returned to India. Unfortunately, both Bogle and the Panchen Lama died in 1781. Hastings despatched a further envoy, Samuel Turner, in 1783. He too was unable to visit Lhasa, despite staying for a period in Tashilumpo. During their journey through Bhutan to Tibet, the medical officer Robert Saunders, writing in the *Philosophical Transactions of the Royal Society* in 1789, made one of the first European observation on the incidence of goitre in the valleys south of the Himalaya.

Tibet now closed her frontiers for many years as the expansionist tendencies of the newly emergent kingdom of Nepal were centred on Lhasa and the Nepalese invaded Tibet. The Tibetans sued for peace in 1789, but Peking suspected England of aiding the Nepalese and closed Tibet. However, in 1811 the eccentric Thomas Manning became the first Englishman to reach Lhasa. He made no serious preparations for the journey and had no support or encouragement of any sort. He had a dubious claim to being a medical practitioner, taking several bottles of fine cherry brandy with him and practising suspect medicine in Lhasa; nevertheless, he gained an audience with the Dalai Lama.

Despite the closure of Tibet, two French Lazarist priests, Fathers Huc and Gabet, entered Lhasa in 1846 to become the first missionaries to see Tibet's capital since the expulsion of the Capuchins a century before. The Abbé Huc's account is full of entertaining detail. He and Gabet were the last Christian mission in Lhasa and stayed only two months.

Because Central Asia was uncharted and closed, and because collision with the eastward-expanding Russians seemed inevitable, it became imperative for the British to gain information about trans-Himalayan regions in order to defend India. Because foreigners would be recognised instantly, the Survey of India, and in particular Montgomerie and Walker, decided to train as explorers Tibetan-speaking members of the communities living south of the Himalaya. Known as the Pandits and identified only by their initials, they joined caravan routes and covered immense distances in inevitably dangerous circumstances, often being absent for many years. They were trained to walk at a uniform 2000 paces to the mile, to count every pace they took, and after 100 paces to drop a bead from their rosaries into their prayer wheels. One Pandit covered 1580 miles (more than 2500km) – that is 3,160,000 paces! They provided information about Lhasa, the Tsangpo gorges, and almost certainly Namjagbarwa and Kula Kangri, the highest peak in Bhutan. 'A.K.' reached the Qaidam basin and east Tibet, and was away for four years. Nain Singh explored western Tibet, specifically the gold-mining area Thok Jhālung – a bleak and desolate plain at nearly 5000m.

Russian geographers who explored Central Asia in the later part of the nineteenth century included Semyenov, who explored the Tien Shan and was the first to classify their flora and fauna. The Fedchenkos concentrated on the Pamir and Trans-Alai (Zaalayskiy Khrebet), while Przevalski worked in the Gobi Desert and north Tibet. Sven Hedin, the formidable Swedish geographer, travelled extensively through the Pamir, Taklimakan and Tibet. He wrote voluminously and made meticulous drawings and maps. Aurel Stein, an archaeologist of great distinction, worked on the 'lost cities' buried in sand north of the Tibetan plateau. He brought back much priceless material, including part of the library from the cave of the thousand Buddhas at Dunhuang in the Gobi Desert, where the world's oldest printed books were found, and greatly increased knowledge of the cultural heritage of Central Asia.

Modern exploration

The exploration of the Himalaya gathered pace in the 1920s. Most expeditions combined science and mountaineering, and the larger peaks attracted specific nationalities. British efforts were concentrated on Everest, which they explored and attempted from the north through Tibet. Eight individuals reached 8535m without oxygen before World War II. Following the opening of Nepal in 1950, the first ascent was made by Hillary and Tensing, using oxygen, in 1953. In 1960 the Chinese made the first ascent from the north, and the first without supplementary oxygen was made by Reinold Messner and Peter Habeler in 1978.

The first ascent of an 8000m peak, Annapurna, was achieved by the French in 1950 and sparked off a series of successful expeditions. K2, the world's second highest peak, was climbed in 1954 by an Italian expedition, and the third highest, Kanchenjunga, by a British party in 1955. Much of the Himalaya has now been mapped and many peaks climbed. The least known is the eastern Himalaya, running along the boundary of Bhutan and the North-East Frontier Agency. Although much of this area has been traversed by the plant-hunters Kingdon Ward, Ludlow and Sherrif, and some of it has been mapped by Bailey and Morshead, no big peaks have been climbed except for Chomolhari, the well-known landmark on the way from Sikkim to Lhasa.

In 1980 the first scientific symposium on the Tibetan plateau was organised in Beijing; it revealed that the Chinese had carried out a considerable amount of work in the previous fifteen years. In addition, a number of peaks in Central Asia were opened to foreigners at this time; these included Everest (north side), Xixaangma (Gosāīnthān),

Maqên Gangri (Amne Machen), Gongga Shan (Minya Konka), Bogdo Ola, Kongur Tobe Feng, Kongur Shan and Muztagata. All had been climbed by either the Chinese or parties from other countries, with the exception of Kongur Shan. A successful British ascent, the first, was made in 1981.

Scientific co-operation in the form of a joint expedition of French and Chinese geologists has been taking place in the last few years. This has concentrated on the structure of the 'Lhasa block' – a region lying between Lhasa and the Himalaya. The structure of central and northern Tibet, and Qinghai, was the object of a joint project undertaken by the Chinese Academy of Sciences and the Royal Society. The traverse from Lhasa to Golmud in the Qaidam basin took place in 1985, and Muztag Feng (Ulugh Mustagh) in the central Kunlun was climbed in the autumn of that year.

Exploration is, of course, leading to many interesting discoveries about the area. Very recently, a new species of cold-tolerant midge has been found living between 5000 and 6000m. This is the first insect known to spend its entire life-cycle in the snow and ice of a glacier, the coldest insect habitat ever recorded. Active at temperatures as low as $-16°C$, the larvae grow in melt-water channels under the ice and feed on blue-green algae and bacteria – a previously unsuspected life chain.

Anecdotal and apocryphal tales abound of lamas living in caves throughout the winter with minimal clothing – or sleeping in the snow without cold injury and of being able to dry wet sheets whilst sitting in the open air by raising their body heat. Some such incidents have been investigated under rigorous scientific conditions, found to be correct, and the findings published in European journals.

Much still remains to be learnt about this inhospitable land and mountain region, the pivot of Asia – an area of increasing scientific and political importance.

Michael Ward
London, January 1987

SELECT BIBLIOGRAPHY

General interest

Allen, C., *A Mountain in Tibet*, André Deutsch, 1982; Futura, 1983.

Aris, M., *Bhutan: the Early History of a Himalayan Kingdom*, Aris & Phillips, 1979.

Avedon, J., *In Exile from the Land of Snows*, Michael Joseph, 1984.

Bailey, F.M., *No Passport to Tibet*, Hart-Davis, 1957.

Bonington, C., *Kongur: China's Elusive Summit*, Hodder & Stoughton, 1982.

Booz, E., *A Guide to Tibet*, Collins, 1986.

Buckley, M., and Strauss, R., *Tibet: a Travel Survival Kit*, Lonely Planet, 1986.

Burdsall, R., and Emmons, A., *Men Against the Clouds: the Conquest of Minya Konka*, John Lane, 1935; Mountaineers, 1980.

Burrard, S.G., and Hayden, H.H., *A Sketch of the Geography and Geology of the Himalaya Mountains and Tibet*, 2nd edn, Government of India, 1933–4.

Cable, M., and French, F., *The Gobi Desert*, Hodder & Stoughton, 1942; Virago Press, 1984.

Cameron, I., *Mountains of the Gods*, Century/Royal Geographical Society, 1984.

Chapman, F.S., *Memoirs of a Mountaineer. Lhasa: the Holy City*, reprinted Sutton, 1984.

Chen, J., *The Sinkiang Story*, Macmillan, 1977.

Cox, E.H.M., *Plant Hunting in China*, Collins, 1945.

Cumming, D., *The Country of the Turkomans*, Royal Geographical Society/Oguz Press, 1977.

Dabbs, J., *History of the Discovery and Exploration of Chinese Turkestan*, Mouton, 1963.

Dalai Lama, *My Land My People*, Weidenfeld & Nicolson, 1962.

De Rachewiltz, I., *Papal Envoys to the Great Khans*, Faber & Faber, 1971.

Earnshaw, G., *On Your Own in China*, Century, 1983.

Fairbank, J.K., (ed.), *The Cambridge History of China – Vol. 10: Late Ch'ing, 1800–1911*, Cambridge University Press, 1978.

Fleming, P., *Bayonets to Lhasa*, Hart-Davis, 1961.

Fleming, P., *News from Tartary*, Jonathan Cape, 1936; Futura, 1980.

Fletcher, H., *A Quest of Flowers*, Edinburgh University Press, 1975.

Fraser-Tyler, W.K., *Afghanistan*, 3rd edn, Oxford University Press, 1967.

Hagen, T., Dyrenfurth, G.O., Schneider, E., and Von Furer-Haimendorf, C., *Mount Everest: Formation, Population and Exploration of the Everest Region*, Oxford University Press, 1963.

Hambly, G., *et al.*, *Central Asia*, Dall/Weidenfeld & Nicolson, 1962.

Harrer, H., *Seven Years in Tibet*, Hart-Davis, 1953.

Harrer, H., *Return to Tibet*, Weidenfeld & Nicolson, 1984.

Hedin, S., *Books out of print but see histories of exploration, and bibliographies.*

Herzog, M., *Annapurna*, Jonathan Cape, 1952.

Hopkirk, P., *Foreign Devils on the Silk Road*, John Murray, 1980.

Hopkirk, P., *Trespassers on the Roof of the World*, John Murray, 1982; Oxford University Press, 1983.

Hopkirk, P., *Setting the East Ablaze*, John Murray, 1984.

Hunt, J., *The Ascent of Everest*, Hodder & Stoughton, 1953.

Jigmei, N.N., *Tibet*, McGraw-Hill, 1981.

Kaplan, F.M., and Keijzer, A.J. de, *The China Guidebook*, 5th edn, Eurasia Press, 1984.

Keay, J., *When Men and Mountains Meet*, John Murray, 1977; Century, 1983.

Keay, J., *The Gilgit Game*, John Murray, 1979.

Kingdon Ward, F., *The Riddle of the Tsangpo Gorges*, Edward Arnold, 1926.

Kish, G., *To the Heart of Asia: the Life of Sven Hedin*, University of Michigan Press, 1984.

Kohli, M.S., *The Himalayas, Playground of the Gods: Trekking, Climbing, Adventure*, Vikas Publishing, 1983.

MacGregor, J., *Tibet: a Chronicle of Exploration*, Routledge & Kegan Paul, 1970.

Maillart, E., *Forbidden Journey: from Peking to Kashmir*, William Heinemann, 1937; Century, 1983.

Maraini, F., *Where Four Worlds Meet: Hindu Kush, 1959*, Hamish Hamilton, 1964.

Mason, K., *Abode of Snow: a History of Himalayan Exploration and Mountaineering*, Hart-Davis, 1955.

Maxwell, N., *India's China War*, Jonathan Cape, 1970.

Migot, A., *Tibetan Marches*, Hart-Davis, 1955.

Miller, K., *Continents in Collision: the International Karakoram Project*, George Philip, 1982.

Mirsky, J., *Great Chinese Travellers*, Allen & Unwin, 1965.

Mirsky, J., *Sir Aurel Stein, Archaeological Explorer*, University of Chicago Press, 1977.

Morgan, G., *Anglo–Russian Rivalry in Central Asia 1810–95*, Cass, 1981.

Myrdal, J., *The Silk Road: a Journey from the High Pamirs and Ili through Sinkiang and Kansu*, Victor Gollancz, 1980.

Peissel, M., *Mustang: a Lost Tibetan Kingdom*, William Collins, 1968.

Prawdin, M., *The Mongol Empire*, Allen & Unwin, 1940.

Richardson, H.E., *Tibet and Its History*, Oxford University Press, 1962.

Rowell, G., *Mountains of the Middle Kingdom*, Sierra Club, 1983; Century, 1985.

Seth, V., *From Heaven Lake*, Chatto & Windus, 1983: Sphere, 1984.

Severin, T., *The Oriental Adventure: Explorers of the East*, Angus & Robertson, 1976.

Shen, Tsung-Lien, and Liu, Shen-Chi, *Tibet and the Tibetans*, Stanford University Press, 1953.

Shipton, E., *The Six Mountain-Travel Books*, Diadem, 1985.

Skrine, C.P., *Chinese Central Asia*, 2nd edn, Methuen, 1971.

Snellgrove, D., *Buddhist Himalaya*, Bruno Cassirer, 1957.

Snellgrove, D., *Himalayan Pilgrimage*, 2nd edn, Routledge & Kegan Paul, 1981.

Snellgrove, D., and Richardson, H.E., *A Cultural History of Tibet*, Weidenfeld & Nicolson, 1968.

Stein, A., *Books out of print but see histories of exploration and bibliographies.*

Stein, R.A., *Tibetan Civilization*, Faber & Faber, 1972.

Swift, H., *Trekker's Guide to the Himalayas and Karakoram*, Hodder & Stoughton, 1982.

Tilman, H.W., *The Seven Mountain-Travel Books*, reprinted Diadem, 1983.

Tregear, T.R., *China: a Geographical Survey*, Hodder & Stoughton, 1980.

Tung, R.R., *A Portrait of Lost Tibet*, Thames & Hudson, 1980
Unsworth, W., *Everest*, Allen Lane, 1981; Penguin, 1982.
Vaurie, C., *Tibet and Its Birds*, H.F. & G., Witherby, 1972.
Waley, A. (trans.), *The Travels of an Alchemist*, Routledge & Kegan Paul, 1931.
Waugh, T., *Marco Polo: a Modern Translation*, Sidgwick & Jackson, 1984.
Woodcock, G., *Into Tibet: the Early British Explorers*, Faber & Faber, 1971.
Younghusband, F., *The Heart of a Continent*, John Murray, 1896.
Zhang, M., *Roof of the World*, Foreign Language Press/Abrams & Beijing, 1982.

Specialist and reference, bibliographies

Baume, L.C., *Silvalaya: the 8000m Peaks of the Himalaya*, Gastons-West Col. 1978.
Bucherer-Dietschi, P., *Bibliotheca Afghanica*, Swchweizerisches Afghanistan-Archiv, 1978–9.
Chi, Jian-Mei, and Ren, Bing-Hui, *Glaciers in China*, Shangai Scientific & Technical Puclications, 1980
Francke, A.H., *Antiquities of Indian Tibet* 2 vols, Government Press, Calcutta, 1914 and 1926.
Gansser, A., *Geology of the Himalaya*, Wiley & Son, 1964.
Gupta, R.K., *Bibliography of the Himalaya*, Indian Documentation Service Gurgon, 1981.
Hayden, Sir H., and Cosson, C., *Sport and Travel in the Highlands of Tibet*, Cobden-Sanderson, 1927.
Hedrick, B.C.E., *et al.*, *A Bibliography of Nepal*, Scarecrow Press, 1973.
Karan, P.K., *Bhutan: a Physical and Cultural Geography*, University of Kentucky Press, 1967.
Karan, P.K., and Jenkins, W.M., *The Himalayan Kingdoms: Bhutan, Sikkim and Nepal*, Van Nostrand, 1963.
Kihara, M. (ed.), *Fauna and Flora of Nepal Himalaya*, Fauna & Flora Research Society, 1955.
Kihara, M. (ed.), *Land and Crops of Nepal Himalaya*, Fauna & Flora Research Society, 1956.
Kihara, M. (ed.), *Peoples of Nepal Himalaya*, Fauna & Flora Research Society, 1957.
Lattimore, O., *Inner Asian Frontiers of China*, (USA Geological Survey Research Series No. 21), US Geological Survey, 1940.

Liu, Dong-Shen, *Mount Tuomar Scientific Expedition (Tien Shan)*, Xinjiang Peoples' Press, 1982.

Markham, Sir C.R., *Narratives of the Mission of George Bogle to Tibet and of the Journey of Thomas Manning to Lhasa*, Trübner, 1876; Manjusri Publishing House, 1971.

Neate, W.R., *Mountaineering and Its Literature*, Cicerone Press, 1978. (Revised edn in progress.)

Polunin, O. and Stainton, A., *Flowers of the Himalaya*, Oxford University Press, 1984.

PROCEEDINGS OF SYMPOSIUM ON QINGHAI-XIZANG (Tibet) PLATEAU, *Geological and Ecological Studies Vol. 1: Geology, Geological History and Origin of Qinghai-Xizang Plateau*, Science Press/Gordon & Breach Science Publishers, 1981.

PROCEEDINGS OF SYMPOSIUM ON QINGHAI-XIZANG (Tibet) PLATEAU, *Geological and Ecological Studies Vol. 2: Environment and Ecology of Qinghai-Plateau*, Science Press/Gordon & Breach Science Publishers, 1981.

Rock, J.F., *The Amnye Ma-chhen Range and Adjacent Regions*, Instituto Italiano per il Medio ed Estremo Oriente, 1956.

Schweinfurth, U., and Schweinfurth-Marby, H., *Exploration in the Eastern Himalayas and the River Gorge Country of Southern Eastern Tibet – Francis (Frank) Kingdom Ward (1885–1958) – an Annotated Bibliography with a Map of the Area of His Expeditions*, Steiner, 1975.

Sherring, C.A., *Western Tibet and the British Borderland*, Edward Arnold, 1906.

Sircar, J. *Himalayan Handbook*, R. Sircan, 1979.

Turner, S., *Account of an Embassy to the Court of Teshoo Lama in Tibet*, Nicol, 1800; Manjusri Publishing House, 1970.

Wissman, H. von, *Die Heutige Vergletscherung und Schneegrenze in Hochasien*, Steiner, 1959.

Yakushi, Y., *Catalogue of the Himalayan Literature*, Yakushi, 1972; Hakusuisha Publishing Co., 1984.

Yoshizawa, I., *et al.*, *Mountaineering Maps of the World No. 1: Great Himalaya*, Gakushuken Kyusua, 1977.

Yoshizawa, I., *et al.*, *Mountaineering Maps of the World No. 2: Karakoram and Hindu Kush*, Gakushken Kyusua, 1978.

NOTE FOR THE READER

This Gazetteer is intended to provide a finding key to places falling within the area of the map. Preferred forms of names are in accordance with the principles and policies of the Permanent Committee on Geographical Names. Because it was considered desirable to give the user the means of locating, approximately, many places for which there was insufficient space on the map itself, a considerable number of additional names has been included in the Gazetteer. An additional reason for this is the recent publication of an official gazetteer of China* which made it possible for the first time to render into the approved Pinyin system of transcription the names of many hundreds of places in the minority-language areas of western China. It was decided to include in the present Gazetteer all place-names occurring in this Chinese publication that fall within the area of the map. Also included is a number of variant and former names and spellings, cross-referenced to the current accepted form. This Gazetteer therefore contains approximately 6400 entries corresponding to more than 5800 different places and features; the number of names on the map itself is approximately 2000.

The listing is arranged in five columns, as follows:

name   In this column there appear, as main entries, the currently
             accepted spellings of names; cross-referenced to these are
             variant and former names. Those names that occur on the
             map itself are distinguished by a preceding asterisk (*).

designation This column gives, in coded form, the designation, or
             definition, of each feature listed in the Gazetteer. The
             designation codes used indicate definitions as follows:

             BDG bridge   CCN   Chinese county name
             BSN basin   (see below for explanation)

*Zhongguo Diminglu, *Zhonghua Renmin Gongheguo Dituji Diming Suoyin* (*Gazetteer of China: An Index to the Atlas of the People's Republic of China*), Ditu Chubanshe (The Map Publishing House), Peking, 1983.

CNL	canal	PK	peak
DEPR	depression	PPL	populated place
DES	desert	⊙PPL	populated place
GLCR	glacier		(see below for
GRGE	gorge		explanation)
ISL	island	RGN	region
LK	lake	RIV	river
MSTY	monastery	RSV	reservoir
MT	mountain	SITE	site
MTS	mountains, mountain	SPNG	spring
	range	VAL	valley
PASS	pass	WELL	well

area   This column gives, in coded form, an indication of the country or area within which each feature is located. Users should note that the presence of a particular code does not necessarily imply any recognition of sovereignty over the feature concerned. The codes used are as follows:

A	Afghanistan	N	Nepal
B	Burma	P	Pakistan
C	China	U	USSR
I	India	X	International Feature
M	Mongolia	Z	Bhutan

lat.   This column gives the latitude north of the equator of each feature listed in the Gazetteer, to the nearest tenth of a degree.

long.   This column gives the longitude east of Greenwich of each feature listed in the Gazetteer, to the nearest tenth of a degree.

The Gazetteer is followed by a supplement, entitled 'Listing of Certain Chinese County and County Centre Names'. This lists alphabetically the names of all Chinese counties (generally *xians*) within the area covered by the map and the settlements that form their centres, where the one differs from the other. Settlements in China that are the centres of counties of a different name are given the designation ⊙PPL in the Gazetteer, corresponding to the ⊙ symbol used to identify them on the map (if plotted). The counties of which they are centres are given the designation CCN. The supplement enables the user to associate the two names.

Very occasionally, the centre of a county cannot be identified precisely with a particular populated place on the information available, and in such cases it is located in the supplement by means of its relevant commune. Apart from these few instances, commune names are not included in this publication.

GLOSSARY OF GENERIC TERMS

The following terms occur as the generic elements of some of the names in this publication. The meanings provided correspond to the specific features found on the map and as main entries in the gazetteer, and therefore may not in every case reflect standard dictionary definitions.

Āb *(Dari)*: river
Beihu: as for Hu
Bhanjyāng *(Nepali)*: pass
Bulag *(Mongol)*: spring
Bulak *(Uighur)*: spring
Bum *(Kachin)*: mountains
Caka *(Tibetan)*: salt lake
Canco: as for Co
Chuan *(Chinese)*: river
Co *(Tibetan)*: lake
Como: as for Co
Coring: as for Co
Cozha: as for Co
Daban *(Tibetan, Uighur)*: pass
Dao *(Chinese)*: island
Dara *(Tajik, Kirgiz)*: river
Darrah *(Pashto)*: river
Daryā *(Dari)*: river
Dar'ya *(Tajik, Kirgiz)*: river
Dashan: as for Shan
Ding *(Chinese)*: peak, mountains
Dolina *(Russian)*: valley
Dongnongchang: as for Nongchang
Feng *(Chinese)*: peak, mountains
Gangri *(Tibetan)*: peak
Ghar *(Pashto)*: mountains,
 mountain range
Gol *(Mongol)*: river
Gongnongqu *(Chinese)*: agro-
 industrial district, equivalent to
 Xian *(q.v.)*
Gongshe *(Chinese)*: commune
Gora *(Russian)*: peak, mountain
Gory *(Russian)*: mountains,
 mountain range
Gou *(Chinese)*: river
Guan *(Chinese)*: pass
Hai *(Chinese)*: lake
He *(Chinese)*: river
Himāl *(Nepali)*: mountain range
Hu *(Chinese)*: lake
Hudag *(Mongol)*: well
Jian *(Chinese)*: peak
Jiang *(Chinese)*: river
Jing *(Chinese)*: well
Kanal *(Russian)*: canal
Kangri *(Tibetan)*: peak, mountains
Kël' *(Kirgiz)*: lake
Khrebet *(Russian)*: mountains,
 mountain range
Kol' *(Kazakh)*: lake
Kosi *(Nepali)*: river
Kowl *(Dari)*: lake
Kowtal *(Dari)*: pass

xxvi

Kuduk *(Uighur)*: spring, well
Kūh *(Dari)*: mountains, mountain range
Kul' *(Kirgiz)*: lake
La *(Tibetan)*: pass
Lednik *(Russian)*: glacier
Lekh *(Nepali)*: mountain range
Lhamco: as for Co
Linchang *(Chinese)*: forest farm
Ling *(Chinese)*: peak, mountains
Monco: as for Co
Muchang *(Chinese)*: pasture, ranch
Namco: as for Co
Nanshan: as for Shan
Nongchang *(Chinese)*: farm
Nur *(Mongol)*: lake
Ozero *(Russian)*: lake
Pendi *(Chinese)*: basin, depression
Pereval *(Russian)*: pass
Pik *(Russian)*: peak
Pünco: as for Co
Qangco: as for Co
Qi *(Chinese)*: banner, equivalent to Xian *(q.v.)*
Qiao *(Chinese)*: bridge
Qu *(Tibetan)*: river
Quan *(Chinese)*: spring
Ri *(Tibetan)*: peak, mountains, ridge
Ri'gyü *(Tibetan)*: mountains
Ringco: as for Co
Rizê *(Tibetan)*: peak, mountain
Say *(Kirgiz)*: river
Shamo *(Chinese)*: desert

Shan *(Chinese)*: mountain(s)
Shankou, Shan-k'ou *(Chinese)*: pass
Shi *(Chinese)*: city with substantial areal extent
Shui *(Chinese)*: river
Shuiku *(Chinese)*: reservoir
Si *(Chinese)*: monastery, temple
Su, Suu *(Kirgiz)*: river
Tag *(Uighur)*: peak, mountain
Tagāb *(Dari)*: river
Tangco: as for Co
Tau *(Kazakh)*: mountain(s)
Too *(Kirgiz)*: mountain(s)
Tso *(Tibetan)*: lake
Uul *(Mongol)*: mountains
Vodokhranilishche *(Russian)*: reservoir
Xia *(Chinese)*: gorge
Xian *(Chinese)*: county
Xinongchang: as for Nongchang
Xongla: as for La
Xubco: as for Co
Xueshan: as for Shan
Yanchi *(Chinese)*: salt lake
Yanhu *(Chinese)*: salt lake
Youqi: as for Qi
Yumco: as for Co
Zangbo *(Tibetan)*: river
Zhen *(Chinese)*: township
Zizhixian *(Chinese)*: autonomous county
Zuoqi: as for Qi

name	designation	area	lat.	long.
Alabuga see: Ala-Buka				
Ala-Buka	RIV	U	41.1N	74.5E
Ala-Buka	PPL	U	41.1N	74.5E
Ala-Buka	RIV	U	41.4N	71.5E
Alag Hu	LK	C	35.5N	97.1E
Ala Gou	RIV	C	42.8N	87.7E
Alakyzrak, Gora see: Gora Alakyzrak				
Alamdo	PPL	C	30.7N	94.2E
Alamedin	RIV	U	42.8N	74.8E
Alatau	PPL	U	43.3N	77.3E
Alay-Ku	PPL	U	40.3N	74.4E
Alayku see: Alay-Kuu				
Alay-Kuu	RIV	U	40.3N	74.4E
Alayskaya Dolina	VAL	U	39.5N	73.0E
*Alayskiy Khrebet	MTS	U	39.7N	72.0E
Alekseyevka	PPL	U	42.7N	70.3E
Alichur	RIV	U	37.8N	73.4E
Alichur	PPL	U	37.8N	73.6E
*Aligarh	PPL	I	27.9N	78.1E
Alimkent	PPL	U	40.9N	69.2E
*Alipur Duar	PPL	I	26.5N	89.6E
Alitangka see: Moincêr				
*Alma-Ata	PPL	U	43.3N	76.9E
Almaaty, Pereval see: Pereval Almaaty				
*Almalyk	PPL	U	40.8N	69.6E
Almaty, Pereval see: Pereval Almaaty				
Almazar see: Zafar				
*Almora	PPL	I	29.6N	79.7E
*Along	PPL	I	28.2N	94.8E
Altenqoke	PPL	C	36.4N	94.8E
*Altun Shan	MTS	C	38.6N	89.0E
*Altun Shan	PK	C	39.3N	93.7E
Altynkan	PPL	U	40.9N	70.8E
Altyn-Topkan	PPL	U	40.6N	69.6E
*Alwar	PPL	I	27.5N	76.6E

name	designation	area	lat.	long.
Alxa Youqi	CCN	C	39.1N	101.7E
Alxa Zuoqi	CCN	C	38.8N	105.7E
Amanbayevo	PPL	U	42.5N	71.2E
Amankol see: Emenkol				
*Ambāla	PPL	I	30.4N	76.8E
Amdo Qu	RIV	C	32.4N	91.7E
Amdo Xian	CCN	C	32.2N	91.6E
Amne Machen see: Maqên Gangri				
Amoikog	PPL	C	32.9N	102.6E
Amqog	PPL	C	34.8N	102.6E
*Amritsar	PPL	I	31.6N	74.9E
Amroha	PPL	I	28.9N	78.4E
Amu Co	LK	C	33.4N	88.7E
*Anantnāg	PPL	I	33.7N	75.2E
Ananʹyevo	PPL	U	42.8N	77.7E
Anbei	PPL	C	40.8N	96.2E
Anbian	PPL	C	28.6N	104.4E
Anbianzhen see: Anbian				
Andarbag	PPL	U	38.2N	71.5E
Andarbak see: Andarbag				
Andir He	RIV	C	37.6N	83.8E
*Andirlangar	PPL	C	37.6N	83.8E
Andizhan	PPL	U	40.7N	72.4E
*Andizhan	PPL	U	40.8N	72.4E
Angcai	PPL	C	33.1N	92.3E
Angdar Co	LK	C	32.7N	89.5E
Anggang	PPL	U	29.5N	90.2E
*Angren	PPL	U	41.0N	70.2E
Anhong	PPL	C	32.5N	103.6E
Anhua	PPL	C	33.5N	105.0E
Anjoman	PPL	A	35.9N	70.4E
Ankyam	PPL	C	32.7N	101.8E
Annanba	PPL	C	39.2N	93.0E
*Annapūrna	PK	N	28.6N	83.8E
Annapūrna Himāl	MTS	N	28.6N	83.9E

3

4

5

name	designation	area	lat.	long.
Banjiegou	PPL	C	43.7N	89.6E
Banli	PPL	C	28.4N	103.9E
Banqiao	PPL	C	25.1N	99.2E
Banqiao	PPL	C	25.1N	103.7E
Banqiao	PPL	C	25.6N	102.9E
Banqiao	PPL	C	29.1N	104.7E
Banqiao	PPL	C	39.2N	100.2E
Bao'an	PPL	C	35.6N	102.0E
Baoding	PPL	C	26.5N	101.6E
Baogunao	PPL	C	26.9N	103.3E
*Baohe	⊙PPL	C	27.1N	99.2E
*Baohe	PPL	C	30.2N	104.7E
Baojishan	PPL	C	36.7N	104.9E
Baomanjie	PPL	C	25.0N	101.7E
*Baoxing	PPL	C	30.4N	102.8E
Baqêm see: Baqên				
Baqên	PPL	C	32.2N	93.4E
Baqên Xian	CCN	C	31.9N	94.0E
Bar	PPL	C	33.7N	79.5E
*Bâra Banki	PPL	I	26.9N	81.2E
Baradêng	PPL	C	32.3N	103.4E
Bâramûla	PPL	I	34.2N	74.4E
Barayqeka	PPL	C	41.2N	86.7E
Barchadiv	PPL	U	38.3N	72.5E
Barchidev see: Barchadiv				
*Bareilly	PPL	I	28.3N	79.4E
Barga	PPL	C	30.8N	81.3E
Bârgâm	PPL	A	35.1N	71.4E
*Barg-e Matâl	PPL	A	35.7N	71.4E
*Bari	PPL	C	30.7N	95.3E
*Barkam	PPL	C	31.9N	102.2E
*Barkol	PPL	C	43.6N	93.0E
*Barkol Hu	LK	C	43.7N	92.8E
Barkol Shan	MTS	C	43.3N	92.6E
*Baroghil Pass	PASS	X	36.9N	73.4E

name	designation	area	lat.	long.
Barong	PPL	C	31.0N	99.3E
Barowghil	PPL	A	36.9N	73.3E
Barowghil, Kowtal-e see: Baroghil Pass				
Barpeta	PPL	I	26.3N	91.0E
Barsalpur	PPL	I	28.2N	72.1E
Barsem	PPL	U	37.6N	71.7E
*Barskoon	PPL	U	42.2N	77.6E
*Bartang	RIV	U	38.2N	72.2E
Bartang	PPL	U	38.1N	71.9E
Bartogay	PPL	U	43.3N	78.5E
Barun	PPL	C	35.9N	97.4E
Barun	PPL	C	36.1N	97.4E
Basar	PPL	C	33.7N	100.6E
Basid	PPL	U	38.2N	72.1E
Baskarkara, Gora see: Gora Baskarkara				
*Basti	PPL	I	26.8N	82.7E
Batâla	PPL	I	31.8N	75.3E
Batan	PPL	C	35.1N	100.6E
*Batang Xian	CCN	C	30.0N	99.1E
Batken	PPL	U	40.1N	70.8E
Batsawul	PPL	A	34.3N	70.9E
Baubashata, Khrebet see: Khrebet Babash-Ata				
Bawolung	PPL	C	28.8N	101.2E
Baxi	PPL	C	33.6N	103.2E
Baxkorgan	PPL	C	39.0N	90.2E
Baxoila Ling	MTS	C	29.2N	97.7E
Baxoi Xian	CCN	C	30.1N	96.9E
Bay see: Baicheng				
Bayan	PPL	C	34.9N	96.7E
*Bayan	⊙PPL	C	36.1N	102.3E
Bayan	PPL	C	36.7N	101.1E
Bayan	PPL	C	42.2N	86.5E
Bayan Borog	PPL	C	39.4N	101.9E
Bayanbulak	PPL	C	43.0N	84.1E
Bayan Gol	RIV	C	37.3N	97.6E

7

8

9

name	designation	area	lat.	long.
Cainnyigoin	PPL	C	32.7N	102.0E
Caiqi	PPL	C	38.2N	102.7E
Cairiwa	PPL	C	34.5N	98.8E
Caka	PPL	C	36.8N	99.0E
Caka	PPL	C	32.5N	82.4E
Caka *see:* Yanhu				
Caka'lho *see:* Yanjing				
*Caka Yanhu	LK	C	36.6N	99.1E
Caluo	PPL	C	29.1N	102.3E
Camco	PPL	C	32.0N	83.6E
Cam Co	LK	C	32.1N	83.5E
Campbellpore *see:* Attock City				
Cangjie	PPL	C	25.6N	101.2E
Caoba	PPL	C	29.9N	103.1E
Cao Daban	PPL	C	38.0N	100.4E
Caohu	PPL	C	41.3N	84.7E
Caojiabu	PPL	C	36.5N	101.9E
Caojian	PPL	C	25.6N	99.1E
Ca Qu	RIV	C	30.0N	83.7E
Ca Qu	RIV	C	34.2N	97.8E
Cawarong	PPL	C	28.5N	98.3E
Cazê	PPL	C	30.0N	86.4E
Cêgnê	PPL	C	32.3N	93.7E
*Cêri	PPL	C	31.2N	85.5E
Cêringgolêb *see:* Dongco				
Cêruma	PPL	C	33.3N	102.1E
Cêrwai	PPL	C	31.5N	97.1E
Cêrzha	PPL	C	31.3N	84.5E
*Cêtar	PPL	C	37.5N	100.4E
Cêwu	PPL	C	29.0N	99.4E
Chabug	LK	C	33.1N	83.2E
Chabulang	PPL	C	28.5N	100.8E
*Chabyêr Caka	LK	C	31.5N	84.0E
Chacang	PPL	C	30.5N	84.7E
Chadak	PPL	U	40.9N	70.8E
Chadian	PPL	C	25.6N	102.3E

name	designation	area	lat.	long.
*Chagdo Kangri	PK	C	34.2N	84.2E
Chaggur	PPL	C	34.2N	94.9E
Chagla	PPL	C	29.8N	87.9E
Chagmar	PPL	C	34.7N	101.6E
Chagna	PPL	C	29.1N	85.8E
Chagyab Xian *see:* Zhag'yab Xian				
Chagyoi	PPL	C	29.2N	98.1E
Cha'gyüngoinba	PPL	C	31.1N	90.6E
Chagzonggoin	PPL	C	30.8N	100.6E
Chahe	PPL	C	26.6N	103.4E
Chainjoin Co	LK	C	35.6N	87.0E
*Chainpur	PPL	N	29.6N	81.2E
Chaiwopu	PPL	C	43.5N	87.9E
*Chakwäl	PPL	P	32.9N	72.9E
*Chala Gou	PPL	C	34.0N	97.6E
Chalaxung	PPL	C	34.2N	97.8E
Chaldovar *see:* Chaldybar				
Chaldybar	PPL	U	42.8N	73.5E
*Chalengkou	PPL	C	38.0N	93.9E
Chali	PPL	C	29.0N	99.0E
*Chaliikong	PPL	C	31.1N	91.5E
Chamalung	PPL	C	36.6N	101.4E
*Chamba	PPL	I	32.6N	76.2E
Chambal	RIV	I	26.4N	77.5E
Chamco	PPL	C	28.5N	87.5E
Chamda	PPL	C	29.9N	91.0E
Chamdo *see:* Qamdo				
*Chäme	PPL	N	28.6N	84.2E
Chamoling	PPL	C	31.3N	96.3E
*Chamu Co	LK	C	34.3N	91.6E
Chanak	PPL	U	42.0N	69.0E
Chandalash	RIV	U	42.0N	71.0E
Chandalashskiy Khrebet	MTS	U	42.0N	71.0E
Chandausi	PPL	I	28.5N	78.7E
*Chandigarh	PPL	I	30.7N	76.9E

11

12

13

name	designation	area	lat.	long.
Dalu	PPL	C	36.3N	104.8E
Dalung	PPL	C	35.5N	100.5E
Damai	PPL	C	35.2N	102.6E
Damanbu	PPL	C	38.8N	100.4E
Damauli	PPL	N	27.9N	84.3E
Damdoi	PPL	C	31.2N	95.3E
Dameigou	PPL	C	37.5N	96.1E
Damiao	PPL	C	37.1N	104.4E
*Damjong	PPL	C	33.6N	95.8E
Damnyain	PPL	C	29.4N	94.6E
Damqog Zangbo see: Maquan He				
*Dam Qu	RIV	C	33.2N	93.5E
*Damquka	⊙PPL	C	30.4N	91.1E
*Damxoi	⊙PPL	C	28.4N	91.4E
Damxung	PPL	C	33.1N	94.3E
Damxung Xian	CCN	C	30.4N	91.1E
Dananchuan Shuiku	RSV	C	36.4N	101.6E
Dananhu	PPL	C	42.5N	93.8E
*Danba	PPL	C	30.9N	101.9E
*Dandeldhura	PPL	N	29.3N	80.6E
Dando	PPL	C	31.2N	101.4E
Dando	PPL	C	31.5N	100.4E
Da'nga	PPL	C	32.5N	103.4E
Dangara	PPL	U	40.6N	70.9E
*Dangchang	PPL	C	34.1N	104.4E
*Dangchengwan	⊙PPL	C	39.5N	94.8E
*Dang He	RIV	C	39.3N	95.5E
*Danghe Nanshan	MTS	C	39.0N	95.5E
Dangjiaxian	PPL	C	35.5N	105.3E
*Dangjin Shankou	PASS	C	39.3N	94.2E
Dang La see: Tanggula Shankou				
Dangla Shan see: Tanggula Shan (both)				
Dangling	PPL	C	31.0N	101.4E
Dángori	PPL	I	27.7N	95.6E
*Dangqên	PPL	C	31.7N	91.8E

name	designation	area	lat.	long.
Dankhar	PPL	I	32.1N	78.2E
Dankova, Pik see: Pik Dankova				
Danleng see: Danling				
Danling	PPL	C	30.0N	103.5E
Danma	PPL	C	37.8N	102.2E
Danshan	PPL	C	30.1N	104.9E
Danshui Hu	LK	C	34.7N	86.3E
Danshui Hu	LK	C	35.4N	88.3E
Daoba	PPL	C	29.0N	100.3E
*Daocheng Xian	CCN	C	29.0N	100.3E
Daojie	PPL	C	24.9N	98.8E
*Daotanghe	PPL	C	36.4N	100.9E
Dapa Dzong see: Daba				
Dapingzi	PPL	C	27.1N	101.2E
Dapuzi	PPL	C	28.1N	101.3E
Da Qaidam see: Da Qaidam Zhen				
Da Qaidam Hu	LK	C	37.8N	95.3E
*Da Qaidam Zhen	PPL	C	37.8N	95.3E
Daqên	PPL	C	31.7N	92.7E
Daqiao	PPL	C	26.6N	102.8E
*Daqiao	PPL	C	26.6N	103.3E
Daqiao	PPL	C	28.7N	102.2E
Daqiao	PPL	C	32.3N	104.3E
Daqiao	PPL	C	33.7N	105.2E
Da Qu	RIV	C	31.2N	95.7E
Da Qu	RIV	C	32.0N	100.0E
Daquan	PPL	C	39.2N	96.8E
Daquan	PPL	C	41.3N	95.2E
Daquanwan	PPL	C	42.7N	93.8E
Darab Co	LK	C	32.4N	83.2E
Daraka	PPL	C	32.5N	95.9E
Darautkurgan see: Daroot-Korgon				
Daräyem, Daryä-ye see: Daryä-ye Daräyem				
Darbaza	PPL	U	41.6N	69.1E
Darbhanga	PPL	I	26.2N	85.9E

name	designation	area	lat.	long.
Darcang	PPL	C	32.8N	100.0E
Dardamty	PPL	U	43.5N	80.1E
Dardo see: Kangding				
Dargo Zangbo	RIV	C	30.7N	86.5E
Darhin Sum	PPL	C	39.8N	104.3E
Dari	PPL	C	32.3N	100.9E
*Därjiling	PPL	I	27.0N	88.3E
Darlag Xian	CCN	C	33.7N	99.7E
Därma Pass	PASS	X	30.5N	80.5E
*Daroot-Korgon	PPL	U	39.6N	72.2E
Dar Qu	RIV	C	33.6N	99.5E
Darrah-ye Katigal	RIV	A	35.5N	71.3E
Darrah-ye Pich	RIV	A	35.0N	70.8E
*Dartang	⊙PPL	C	31.9N	94.0E
Daru Co	LK	C	31.6N	90.6E
Darvazskiy Khrebet	MTS	U	38.5N	71.3E
Daryā-ye Āq Sū	RIV	A	37.3N	74.3E
Daryā-ye Darāyem	RIV	A	37.0N	70.2E
*Daryā-ye Konar	RIV	A	35.0N	71.5E
*Daryā-ye Kowkcheh	RIV	A	37.1N	70.5E
*Daryā-ye Panj	RIV	A	36.9N	71.5E
*Daryā-ye Ravenj Āb	RIV	A	37.8N	70.3E
Daryā-ye Shiveh	RIV	A	37.5N	71.0E
Daryā-ye Vākhjīr	RIV	A	37.0N	74.2E
Daryā-ye Vardūj	RIV	A	36.5N	71.3E
Darzhuo	PPL	C	32.1N	89.0E
Dashanbao	PPL	C	27.4N	103.2E
Dashi	PPL	C	32.3N	104.9E
Dashigou	PPL	C	31.8N	101.1E
Dashipu	PPL	C	25.0N	101.4E
Dashsitou	PPL	C	43.7N	91.2E
Dashtidzhum	PPL	U	38.0N	70.2E
Dashu	PPL	C	29.3N	102.6E
Dashujing	PPL	C	27.0N	103.5E
Dashuiqiao	PPL	C	36.6N	99.4E
Dashuitou	PPL	C	36.7N	104.8E
*Dasongshu	PPL	C	26.2N	102.5E
Da Surmang	PPL	C	32.5N	96.8E
Datan	PPL	C	38.7N	103.2E
Datarma	PPL	C	32.2N	99.7E
Datian	PPL	C	26.3N	101.7E
Datong	PPL	C	36.6N	103.3E
*Datong He	RIV	C	36.9N	102.7E
*Datong Shan	MTS	X	37.8N	99.9E
Datong Xian	CCN	C	36.9N	101.6E
Datouyang	PPL	C	37.7N	95.6E
*Daulpur	PPL	I	26.7N	77.9E
*Daūsa	PPL	I	26.8N	76.4E
Dawa Co	LK	C	31.2N	84.9E
Dawan	PPL	C	27.8N	103.8E
Dawanqi	PPL	C	40.5N	99.7E
Dawanzi	PPL	C	41.7N	81.4E
Dawatang	PPL	C	25.9N	101.8E
Dawaxung	PPL	C	31.5N	93.1E
Dawê	PPL	C	30.9N	102.6E
Dawê	PPL	C	32.1N	101.7E
Dawo see: Dawu				
*Dawu	⊙PPL	C	34.5N	100.2E
*Dawu	PPL	C	31.0N	101.3E
*Dawukou	⊙PPL	C	39.0N	106.3E
Dawusi	PPL	C	37.8N	91.1E
Daxia	PPL	C	36.5N	102.2E
*Daxia He	RIV	C	35.5N	103.0E
*Daxiang Ling	MTS	C	29.5N	102.7E
*Daxihai Shuiku	RSV	C	40.6N	87.5E
Daxing	PPL	C	26.6N	101.4E
*Daxing	⊙PPL	C	27.2N	100.8E
Daxing	PPL	C	27.7N	103.4E
Daxingdi	PPL	C	26.0N	98.9E

name	designation	area	lat.	long.
Daxue Shan	MTS	C	30.3N	101.8E
*Daxue Shan	MTS	C	39.6N	96.2E
*Dayan	⊙PPL	C	26.8N	100.2E
Dayao Xian	CCN	C	25.7N	101.3E
Dayekou	PPL	C	38.5N	100.2E
Dayêr	PPL	C	31.1N	97.1E
Dayi	PPL	C	30.5N	103.5E
Daying	PPL	C	25.7N	100.4E
Da Yultuz	PPL	C	41.5N	82.4E
Dazhai	PPL	C	27.2N	102.9E
Dazhai	PPL	C	27.5N	103.2E
Dazhuang	PPL	C	26.7N	101.5E
Dazigou	PPL	C	43.4N	93.9E
Dêca	PPL	C	29.6N	100.7E
*Dechang	PPL	C	27.4N	102.1E
Dêdartang	PPL	C	30.2N	99.4E
Dêdêmêl'	PPL	U	41.4N	74.4E
*Degâna	PPL	I	26.8N	74.3E
Dêgbu	PPL	C	29.0N	100.8E
Dêgên see: Dêqên				
Degeres	PPL	U	43.2N	75.8E
*Dêgê Xian	CCN	C	31.8N	98.6E
Dê'gyi	PPL	C	28.0N	87.6E
Dehenglong	PPL	C	35.9N	102.1E
Deh Gholâmân	PPL	A	37.0N	73.1E
*Dehra Dun	PPL	I	30.3N	78.1E
Deju	PPL	C	25.0N	100.6E
Deka see: Pênda				
Dêlêg	PPL	C	29.9N	87.6E
*Delhi	PPL	I	28.6N	77.1E
Deliechuka	PPL	C	33.9N	92.6E
*Delingha	PPL	C	37.3N	97.3E
Delingha Nongchang	PPL	C	37.3N	97.2E
*Demchok	PPL	I	32.7N	79.5E
*Dêmo La	PASS	C	29.3N	97.0E
Dêmqog see: Demchok				
Dênai see: Garyi				
Dengchuan	PPL	C	25.9N	100.1E
Dênggar	PPL	C	29.2N	85.8E
Dengguanzhen	PPL	C	29.1N	104.9E
*Dêngkagoin	⊙PPL	C	34.0N	103.2E
Dêngla Ri'gyü	MTS	C	31.0N	82.4E
Dêngqên Xian	CCN	C	31.5N	95.6E
Dengshen	PPL	C	30.8N	102.9E
Dengyingyan	PPL	C	29.8N	104.7E
*Dêngzê	PPL	C	32.8N	81.4E
Dêntala see: Dên-Talaa				
Dên-Talaa	PPL	U	42.1N	76.6E
Deoria	PPL	I	26.5N	83.8E
*Deosai Plains	RGN	P	35.0N	75.4E
*Deo Tibba	PK	I	32.2N	77.4E
Depsang Plains	RGN	I	35.0N	78.0E
*Dêqên	⊙PPL	C	29.6N	91.4E
Dêqên	PPL	C	29.9N	90.7E
*Dêqên	PPL	C	30.5N	90.1E
*Dêqên Xian	CCN	C	28.5N	98.9E
Dêrdoin	PPL	C	31.9N	98.0E
Dêrdoin	PPL	C	32.0N	97.0E
Dêrlagsumdo	PPL	C	33.8N	96.4E
Dêrmang	PPL	C	33.4N	100.1E
*Dêrong Xian	CCN	C	28.7N	99.3E
Dêr Qu	RIV	C	34.0N	96.5E
Derstei	PPL	C	40.5N	101.4E
Dêrtang	PPL	C	30.3N	95.3E
Dêrub	PPL	C	33.3N	79.7E
Detuo	PPL	C	29.5N	102.2E
*Devli	PPL	I	25.8N	75.4E
*Dewangiri	PPL	Z	26.9N	91.5E
Dêxin see: Kaimar			28.6N	86.6E
Dêxing	PPL	C	29.3N	95.2E

name	designation	area	lat.	long.
Dowa	PPL	C	35.3N	101.9E
Dowa	PPL	C	36.6N	101.5E
Dowr Bābā	PPL	A	34.1N	70.9E
Doxoggu	PPL	C	31.7N	90.1E
Doxong see: Paiqu				
Dozê see: Zangsar				
*Drās	PPL	I	34.4N	75.8E
Drungjiang	PPL	C	27.7N	98.3E
*Drung Jiang	RIV	C	28.3N	98.3E
Druzhba	PPL	U	43.1N	76.9E
*Dŭdu	PPL	I	26.7N	75.3E
Dugh Ghalaṭ	PPL	A	37.1N	70.4E
Duiyan	PPL	C	29.9N	102.9E
Dujiang Yan	PPL	C	30.9N	103.5E
*Dukou	PPL	C	26.5N	101.7E
Dulansi	PPL	C	37.0N	98.6E
Dulan Xian	CCN	C	36.3N	98.1E
*Dulishi Hu	LK	C	34.7N	81.8E
*Dumre	PPL	N	27.9N	84.4E
*Dunai	PPL	N	29.0N	82.9E
Dundwa Range	MTS	N	27.7N	82.5E
Dung Büree see: Dongbolhai Shan				
*Dung Co	LK	C	31.7N	91.1E
*Dunhuang	PPL	C	40.1N	94.6E
Dunxu	PPL	C	28.2N	91.9E
Duokake	PPL	C	36.6N	92.6E
Duomei	PPL	C	26.3N	100.3E
Duomula	PPL	C	34.3N	82.5E
Duonangdongzai Qu	RIV	C	35.4N	95.8E
Duo Qu see: Do Qu				
Duoyue	PPL	C	30.1N	103.7E
Düräj	PPL	A	37.9N	70.7E
Düri	PPL	C	30.9N	97.5E
Dur'ngoi	PPL	C	34.4N	100.1E
Dushanzi	PPL	C	39.6N	94.4E

name	designation	area	lat.	long.
Dushokha, Gora see: Gora Dushokha				
Dusong	PPL	C	31.2N	101.9E
Duwa	PPL	C	37.1N	79.0E
Düxanbibazar	PPL	C	37.6N	80.3E
Dyurbel'dzhin see: Bayetovo				
Dzhail'gan see: Dzhayyilgan				
Dzhakhanabad	PPL	U	41.7N	70.0E
Dzhalal-Abad	PPL	U	40.9N	73.0E
Dzhaldzhir	RIV	U	41.3N	76.8E
Dzhamantau, Khrebet see: Gory Dzhaman-Too				
Dzhaman-Too, Gory see: Gory Dzhaman-Too				
Dzhambul	PPL	U	42.7N	80.0E
*Dzhambul	PPL	U	42.9N	71.4E
*Dzhany-Bazar	PPL	U	41.7N	70.9E
Dzhany-Pakhta	PPL	U	43.1N	74.3E
Dzhavshangoz	PPL	U	37.3N	72.4E
*Dzhayyilgan	PPL	U	39.3N	71.5E
Dzhergalan	PPL	U	42.6N	79.0E
Dzher-Këchkyu	PPL	U	41.9N	76.3E
Dzhetim, Khrebet see: Khrebet Dzhetim				
Dzhetim-Bel', Khrebet see: Khrebet Dzhetim-Bel'				
Dzhety-Oguz	PPL	U	42.3N	78.2E
Dzhety-Oguz	PPL	U	42.4N	78.2E
Dzhirgatal'	PPL	U	39.2N	71.2E
Dzhoon-Aryk	RIV	U	42.0N	75.7E
Dzhuku see: Dzhuuku				
Dzhumashuy	PPL	U	40.8N	71.5E
Dzhumgal	RIV	U	41.9N	74.5E
Dzhumgaltau, Khrebet see: Khrebet Dzhumgal-Too				
Dhumgal-Too, Khrebet see: Khrebet Dzhumgal-Too				
Dzhuuku	RIV	U	42.2N	77.9E
Dzhuvanaryk see: Dzhoon-Aryk				
Dzhylamysh	RIV	U	42.7N	74.4E
*Ebian	PPL	C	29.2N	103.6E
*Ehen Hudag	⊙PPL	C	39.1N	101.7E

name	designation	area	lat.	long.
*Fulin	⊙PPL	C	29.3N	102.6E
Fulu	PPL	C	29.3N	103.6E
Fuxing	PPL	C	31.9N	104.5E
*Gabasumdo	⊙PPL	C	35.2N	100.5E
Gadêgoinba	PPL	C	37.7N	100.5E
Gadê Xian	CCN	C	33.9N	99.9E
Gadong	PPL	C	29.2N	89.1E
Gafurov	PPL	U	40.2N	69.7E
Gahai	PPL	C	34.2N	102.3E
Ga Hai	LK	C	37.0N	97.5E
Ga Hai	LK	C	37.0N	100.6E
Gahai	PPL	C	37.2N	97.6E
Gahe	PPL	C	38.8N	97.7E
Gaidain'goinba	PPL	C	37.2N	99.8E
*Gäighät	PPL	N	26.8N	86.7E
Gaisanggoin	PPL	C	33.3N	97.0E
Gaka	PPL	C	30.8N	101.2E
*Gakuch	PPL	P	36.2N	73.8E
*Gala	PPL	C	28.2N	89.3E
Gala	PPL	C	28.3N	101.3E
Gamba	PPL	C	28.2N	88.4E
*Gamba	PPL	C	28.2N	88.5E
*Gamda	⊙PPL	C	32.3N	100.9E
Gamqên	PPL	C	30.8N	95.1E
Gamtog	PPL	C	31.6N	98.5E
Gana	PPL	C	32.0N	101.0E
Ganai	PPL	C	31.8N	101.5E
Ganben	PPL	C	26.3N	98.9E
Gancaodian	PPL	C	35.8N	104.3E
Gancaogou see: Gancaohu				
Gancaohu	PPL	C	41.7N	88.4E
*Ganchansi	PPL	C	37.0N	102.3E
Gancheng	PPL	C	37.2N	103.4E
Ganda	PPL	C	31.5N	94.2E
*Gandak	RIV	I	26.8N	84.2E

name	designation	area	lat.	long.
Ganding	PPL	C	25.3N	98.8E
Gando see: Gandu				
Gandu	PPL	C	35.9N	102.2E
*Ganesh	PK	N	28.4N	85.1E
*Ganga	RIV	I	26.1N	80.8E
*Gangānagar	PPL	I	29.9N	73.9E
Gangca	PPL	C	33.7N	96.1E
Gangca Dasi	PPL	C	37.5N	100.1E
Gangca Xian	CCN	C	37.3N	100.2E
Gangca Zhan	PPL	C	37.2N	100.1E
*Gangdisê Shan	MTS	C	30.6N	83.0E
Gangfang	PPL	C	26.1N	98.5E
Gangga	PPL	C	30.4N	93.9E
Gangmar Co	LK	C	33.8N	84.3E
Gangnyi	PPL	C	32.3N	90.7E
Gangou	PPL	C	26.6N	103.3E
Gangou	PPL	C	35.4N	105.5E
Gangou	PPL	C	38.9N	97.1E
Gangouyi	PPL	C	35.9N	105.0E
*Gangtok	PPL	I	27.3N	88.7E
*Gangu	PPL	C	34.7N	105.3E
Gan'gyur	PPL	C	37.2N	98.9E
Ganjia	PPL	C	35.4N	102.5E
Ganjiang	PPL	C	29.7N	103.6E
Ganjig	PPL	C	37.1N	100.5E
Ganjunpu	PPL	C	38.9N	100.1E
Ganluchi	PPL	C	36.7N	103.7E
Ganluo Xian	CCN	C	28.9N	102.7E
*Ganq	PPL	C	37.3N	92.5E
Ganquan	PPL	C	33.5N	105.1E
Ganquan	PPL	C	34.4N	105.9E
Gansenquan Hu	LK	C	37.4N	92.8E
Gantang	PPL	C	37.4N	104.5E
*Ganxiangying	⊙PPL	C	28.3N	102.4E
*Ganzhou	⊙PPL	C	38.9N	100.4E

22

name	designation	area	lat.	long.
Gaoba	PPL	C	37.8N	102.6E
Gaochang	PPL	C	28.8N	104.3E
Gaochang Gucheng	SITE	C	42.8N	89.6E
Gaojiadun	PPL	C	37.4N	103.8E
Gaolan Xian	CCN	C	36.3N	103.9E
*Gaoligong Shan	MTS	X	26.9N	98.7E
Gaomiao	PPL	C	29.6N	103.2E
Gaomiao	PPL	C	36.4N	102.0E
Gaopian	PPL	C	31.0N	104.2E
Gaoqiao	PPL	C	25.6N	102.2E
Gaoqiao	PPL	C	28.0N	103.8E
Gaoqiao	PPL	C	34.1N	105.9E
*Gaotai	PPL	C	39.3N	99.8E
Gaotaizhan	PPL	C	39.2N	99.7E
Gaoya	PPL	C	35.7N	104.2E
Gaozhuoying	PPL	C	28.6N	103.4E
Gaparma	PPL	C	33.3N	102.4E
Gaqag	PPL	C	28.2N	100.1E
Gaqoi	PPL	C	29.9N	83.5E
Gaqung see: Surco				
*Gar	PPL	C	32.1N	80.0E
Garang	PPL	C	36.2N	101.5E
Garangsi	PPL	C	35.2N	101.9E
Garbiutangka	PPL	C	32.0N	80.1E
*Garbo	⊙PPL	C	28.4N	90.8E
Garco	PPL	C	33.3N	88.5E
Gargu Yan	PPL	C	33.8N	92.3E
Gariqiong	PPL	C	29.9N	90.3E
Garkung Caka	LK	C	33.9N	86.5E
*Garkyagdêugang	PK	C	33.4N	90.8E
Garla Lhamco	LK	C	34.4N	97.7E
Garm	PPL	U	39.0N	70.4E
Garma	PPL	C	31.8N	96.9E
Garmo, Pik see: Pik Garmo				
Garmoyangkyi	PPL	C	35.6N	100.2E

name	designation	area	lat.	long.
Garong	PPL	C	31.4N	93.6E
*Gar Qu	RIV	C	33.0N	102.5E
*Gar Qu	RIV	C	33.8N	91.8E
Garrong	PPL	C	28.4N	100.5E
Gartar see: Qianning				
*Gartog	⊙PPL	C	29.6N	98.5E
Gartok see: Garyarsa				
Garyarsa	PPL	C	31.8N	80.4E
Garyi	PPL	C	30.8N	99.0E
*Gar Zangbo	RIV	C	32.2N	80.0E
*Garzê	PPL	C	31.6N	99.9E
Gase	PPL	C	31.9N	86.2E
*Gasherbrum I	PK	X	35.7N	76.7E
Gasherbrum II	PK	X	35.8N	76.7E
Gas Hu see: Gas Hure Hu				
*Gas Hure Hu	LK	C	38.1N	90.9E
Gatang	PPL	C	30.1N	93.3E
*Gauhâti	PPL	I	26.2N	91.8E
*Gaur	PPL	N	26.8N	85.3E
Gava	PPL	U	41.1N	71.1E
Gava-Say	RIV	U	41.2N	71.0E
Gaxiu	PPL	C	34.4N	102.3E
Gaxiugou	PPL	C	36.6N	98.8E
Gaxung	PPL	C	32.0N	95.7E
*Gaxun Nur	LK	C	42.4N	100.7E
Gayahe	PPL	C	36.3N	95.3E
Gaz	PPL	U	39.8N	71.0E
Gazalkent	PPL	U	41.6N	69.8E
*Gêding	PPL	C	29.2N	88.3E
*Gêding	⊙PPL	C	29.4N	88.2E
Gêgar	PPL	C	31.5N	81.8E
Gê'gyai Xian	CCN	C	32.4N	81.1E
Gêla	PPL	C	30.2N	83.5E
*Gêladaindong	PK	C	33.5N	91.0E
Geliping	PPL	C	26.6N	101.5E

34

name	designation	area	lat.	long.
Jongnê	PPL	C	30.7N	96.8E
*Jorhät	PPL	I	26.8N	94.2E
Jor Hu	LK	C	39.5N	79.0E
*Jorm	PPL	A	36.9N	70.9E
Jorra	PPL	C	28.2N	92.3E
*Joshimath	PPL	I	30.6N	79.6E
Judian	PPL	C	27.2N	99.6E
Juexizhen	PPL	C	28.9N	104.2E
Juh He	RIV	C	36.6N	94.0E
*Juhongtu	PPL	C	37.9N	96.3E
Jujiaji	PPL	C	35.6N	102.9E
Jullundur	PPL	I	31.3N	75.7E
Julong	PPL	C	28.8N	100.5E
Jumanggoin	PPL	C	32.4N	98.5E
*Jumla	PPL	N	29.3N	82.2E
*Jumlikhalangā	PPL	N	28.6N	82.5E
Junbesi	PPL	N	27.6N	86.5E
Jungba	PPL	C	31.6N	94.2E
Jungsi	PPL	C	32.5N	90.0E
Jun Mahai Hu	LK	C	38.2N	94.3E
Jun Ul	PPL	C	37.4N	97.5E
*Jun Ul Shan	MTS	C	37.6N	97.3E
Ju'nyung	PPL	C	33.0N	98.2E
Ju'nyunggoin see: Ju'nyung				
*Jurhen Ul Shan	MTS	C	34.0N	91.0E
Jushike	PPL	C	36.5N	99.8E
Juyan	PPL	C	42.3N	100.8E
Jyekundo see: Gyêgu				
*K2	PK	X	35.9N	76.5E
*Kābul	RIV	P	34.2N	71.4E
Kabutiyên	PPL	U	38.9N	70.0E
Kachik-Alay, Khrebet see: Khrebet Kachik-Alay				
*Kada	PPL	C	28.1N	87.3E
Kadzhi-Say	PPL	U	42.1N	77.2E
*Kāgān	PPL	P	34.8N	73.5E

name	designation	area	lat.	long.
Kagang	PPL	C	35.3N	100.2E
Kagor	PPL	C	31.1N	88.4E
Kaiba	PPL	C	36.9N	98.5E
Kaibamardang	PPL	C	38.8N	97.6E
*Kaidu He	RIV	C	42.4N	85.3E
Kaika	PPL	C	29.2N	87.1E
Kailās see: Kangrinboqê Feng				
Kailās Range see: Gangdisê Shan				
Kaimar	PPL	C	28.6N	86.6E
Kaimar	PPL	C	31.1N	85.7E
Kaimar	PPL	C	31.1N	87.3E
Kaimar	PPL	C	33.1N	95.8E
Kaimo	PPL	C	29.3N	100.4E
Kainda see: Kaindy				
Kaindy	PPL	U	42.9N	73.7E
Kaindy-Katta, Khrebet see: Khrebet Kayyngdy-Katta				
Kaiwen	PPL	C	27.1N	99.8E
Kākān	PPL	A	37.2N	70.3E
Kakshaal-Too, Khrebet see: Khrebet Kakshaal-Too				
*Kālābāgh	PPL	P	33.0N	71.6E
*Kalaikhumb	PPL	U	38.5N	70.8E
Kalaiya	PPL	N	27.0N	85.0E
*Kalām	PPL	P	35.5N	72.6E
Kalān Eylgah	PPL	A	37.4N	70.7E
Kalat	PPL	A	36.0N	70.6E
Kāli see: Mahākāli				
Kāli Gandaki	RIV	N	27.8N	83.6E
*Kālimpang	PPL	I	27.0N	88.6E
Kalinin	PPL	U	41.3N	69.2E
Kalininskoye	PPL	U	42.8N	73.8E
Kalinovka	PPL	U	42.7N	75.4E
Kalmakkuduk	PPL	C	37.0N	82.8E
Kalpa	PPL	I	31.6N	78.3E
Kalpaktobe	PPL	U	42.9N	71.4E
*Kalpin	PPL	C	40.5N	79.0E

name	designation	area	lat.	long.
Kaltila	PPL	C	38.2N	77.0E
Kamba Dzong *see:* Gamba				
Kambardi	PPL	C	28.2N	88.5E
Kamberdi *see:* Kambardi				
Kambudoiqên	MTS	C	40.1N	80.2E
*Kámdeysh	PPL	A	35.4N	71.3E
Kámeh	PPL	A	34.4N	70.6E
Kamenka	PPL	U	42.9N	72.8E
*Kámet	PK	I	30.9N	79.6E
Kamlung	PPL	C	38.8N	99.9E
Kam'ong	PPL	C	28.1N	101.0E
Kamru	PPL	C	28.6N	89.4E
Kamsa	PPL	A	29.4N	96.8E
Kámü	PK	X	35.4N	71.4E
*Kánchenjunga	PPL	C	27.7N	88.2E
Ka'ngai *see:* Kata				
Kangchenjunga *see:* Kánchenjunga				
*Kangding	PPL	C	30.0N	101.9E
*Kanggar	PPL	C	28.6N	100.3E
Kangardo Rizê *see:* Kangto				
Kangkir	PPL	C	37.1N	78.3E
Kangle	PPL	C	35.3N	103.6E
Kanglong *see:* Sangruma				
Kanglung	PPL	C	31.4N	87.5E
Kanglung	PPL	C	33.8N	100.2E
*Kangmar	PPL	C	28.5N	89.6E
Kangmar	PPL	C	30.6N	85.6E
Kangqung Kangri	MTS	C	30.5N	84.6E
Kángra	PPL	I	32.1N	76.3E
Kangrai	PPL	C	32.0N	90.5E
*Kangri Bolhug	PK	C	32.5N	87.9E
Kangrigarbo Qu	RIV	C	28.9N	96.6E
*Kangrinboqê Feng	PK	C	31.0N	81.3E
Kangro	PPL	C	33.0N	85.5E
Kangsê	PPL	C	31.5N	96.2E
*Kangto	PK	X	27.8N	92.4E
Kangtog	PPL	C	32.5N	84.3E
Kangxiwar	PPL	C	36.2N	78.7E
Kangxung	PPL	C	29.2N	90.1E
*Kangzhag Ri	PK	C	35.5N	89.5E
*Kanibadam	PPL	U	40.3N	70.4E
*Kánjiroba	PK	N	29.4N	82.7E
*Kánpur	PPL	I	26.5N	80.2E
Kansay	PPL	U	40.5N	69.7E
Kansu	PPL	C	39.7N	75.0E
*Kant	PPL	U	42.9N	74.9E
Kántiwa	PPL	A	35.3N	70.7E
Kapkatas, Khrebet *see:* Khrebet Kapka-Tash				
Kapka-Tash, Khrebet *see:* Khrebet Kapka-Tash				
Ka Qu	RIV	C	31.2N	95.2E
Kaqung	PPL	C	37.9N	76.8E
*Karaart	PPL	U	39.1N	73.6E
Karabagish *see:* Sovetabad				
*Kara-Balta	RIV	U	42.8N	73.9E
Kara-Balto	RIV	U	42.6N	73.8E
Kara-Balty *see:* Kara-Balta				
Karabalty *see:* Kara-Balto				
Karabulak	PPL	C	40.7N	77.5E
Karabulak	PPL	U	42.5N	69.8E
Kara-Buura	RIV	U	42.4N	71.5E
Karadar'ya	RIV	U	40.9N	72.0E
Kara-Dzhorgo, Khrebet *see:* Khrebet Kara-Dzhorgo				
Karajül	PPL	C	40.1N	76.7E
Kara-Kastek	RIV	U	43.1N	76.1E
Kara-Katty, Khrebet *see:* Khrebet Kara-Katty				
Karakax *see:* Moyu				
*Karakax He	RIV	C	36.6N	78.8E
Karakemer	PPL	U	43.1N	71.0E
Karakhitay	PPL	U	40.9N	69.7E

name	designation	area	lat.	long.
Kogart, Pereval *see:* Pereval Kogart				
Kogbo	PPL	C	32.6N	101.5E
Kogershin	PPL	U	42.8N	72.5E
*Kohāt	PPL	P	33.6N	71.4E
*Koikyim Qu	RIV	C	34.4N	94.7E
*Kokand	PPL	U	40.5N	70.9E
Kokankishlak *see:* Pakhtaabad				
Kokkoynak	PPL	C	41.0N	85.7E
Koko Nor *see:* Qinghai Hu				
Koko-Shili Range *see:* Hoh Xil Shan				
Kokpek	PPL	U	43.5N	78.7E
Koksay	PPL	U	43.0N	79.9E
Kokshaal–Tau, Khrebet *see:* Khrebet Kakshaal-Too				
Koktal	PPL	C	39.4N	74.1E
Kok-Tash	PPL	U	41.2N	72.4E
Kokterak	PPL	C	37.7N	78.2E
Kokterak	PPL	C	42.9N	81.6E
Kokterek	PPL	U	41.3N	69.0E
Kokuybel'	RIV	U	38.8N	73.2E
*Kok-Yangak	PPL	U	41.0N	73.2E
Kokyar	PPL	C	36.8N	81.5E
Kokyar	PPL	C	37.4N	77.1E
*Kol'zhat	PPL	U	43.5N	80.7E
Kommunizma, Pik *see:* Pik Kommunizma				
*Komsomolabad	PPL	U	38.8N	69.9E
Komsomol'skiy	PPL	U	40.4N	71.8E
Konar, Daryā-ye *see:* Daryā-ye Konar				
Konglu	PPL	B	27.3N	97.9E
Kongtan	PPL	C	29.1N	104.6E
*Kongur Shan	PK		38.6N	75.3E
Kongur Tobe Feng	PK	C	38.6N	75.1E
Kongzakug	PPL	C	32.3N	97.0E
*Konqi He	RIV	C	40.8N	87.8E
Koram	PPL	U	43.6N	78.2E
Koramlik	PPL	C	37.4N	85.8E

name	designation	area	lat.	long.
Korān va Monjān	PPL	A	36.0N	70.8E
Korgan	PPL	C	37.1N	85.0E
*Korla	PPL	C	41.7N	86.1E
Korlondo	PPL	C	31.9N	98.6E
*Kormang	PPL	C	31.7N	92.4E
Kornilovka	PPL	U	42.6N	70.2E
Korqên	PPL	C	30.6N	96.7E
Korqug	PPL	C	30.4N	89.7E
Korra	PPL	U	30.8N	97.8E
Kosh-Dëbë	PPL	U	41.1N	74.3E
Koshtëbë *see:* Kosh-Dëbë				
Kosrap	PPL	C	37.9N	76.2E
*Kotli	PPL	P	33.5N	73.9E
Koumenzi	PPL	C	43.3N	93.6E
Kowkcheh, Daryā-ye *see:* Daryā-ye Kowkcheh				
Kowl-e Shiveh	LK	A	37.4N	71.3E
Kowtal-e Barowghil *see:* Baroghil Pass				
Kowtal-e Do Rāh *see:* Dorah Pass				
Kowtal-e Khaybar *see:* Khyber Pass				
Kowtal-e Vākhjir *see:* Wakhjir Pass				
Kox Kuduk	WELL	C	40.4N	92.5E
Koxlangza	PPL	C	39.9N	90.7E
*Koxlax	PPL	C	38.0N	80.5E
Koxtag	PPL	C	37.4N	78.0E
Koytezek, Pereval *see:* Pereval Koytezek				
Krasnogorka	PPL	U	43.3N	75.2E
Krasnogorskiy	PPL	U	41.1N	69.8E
Krasnooktyabr'skiy *see:* Shopokov				
Kuangshanzhen	PPL	C	26.6N	103.6E
Küba	PPL	C	33.1N	80.0E
Kucha *see:* Kuqa				
Kuchāman	PPL	I	27.2N	74.9E
Kuchkak	PPL	U	40.3N	70.3E
Küda	PPL	C	36.7N	76.9E
Kudara	RIV	U	38.3N	72.5E

name	designation	area	lat.	long.
Kudara	PPL	U	38.4N	72.7E
Kugart see: Kêk-Art				
Kugka Lhai	PPL	C	31.9N	88.3E
Kugri	PPL	C	35.9N	98.0E
Ku Hai	LK	C	35.2N	99.1E
Kūh-e Khvājeh Moḥammad	MTS	A	36.4N	70.3E
Kūh-e Safid Khers	MTS	A	37.8N	70.9E
*Kula Kangri	PK	C	28.3N	90.6E
Kulanak	PPL	U	41.4N	75.5E
Kulansarak see: Hulangshan				
Kulin, Gora see: Gora Kulin				
*Kulu	PPL	I	31.9N	77.1E
Kumbhakarna Himāl	MTS	N	27.7N	87.2E
Kum Kuduk	WELL	C	40.2N	91.8E
*Kumon Range	MTS	B	26.5N	97.3E
Kumtag Shamo	DES	C	39.7N	92.0E
Kumul see: Hami				
Kümüx	PPL	C	42.2N	88.1E
Kumyshtag see: Kyumyush-Tak				
Kumyshtag, Pik see: Pik Kyumyush-Tak				
Künas see: Künes Linchang				
Künas Chang see: Künes Chang				
Künes see: Xinyuan				
Künes Chang	PPL	C	43.4N	83.2E
Künes Linchang	PPL	C	43.6N	82.6E
Künes Linchang	PPL	C	43.2N	84.6E
Kungey-Alatau, Khrebet see: Khrebet Kyungëy-Ala-Too				
*Kunggar	⊙PPL	C	29.8N	91.7E
*Kunggyü Co	LK	C	30.6N	82.1E
Kunglung	PPL	C	30.4N	86.0E
Kungtang	PPL	C	28.8N	84.7E
Kunjirap	PPL	C	36.9N	75.5E
Kunjirap Daban see: Khūnjerāb Pass				
*Kunlun Shan	MTS	C	36.0N	84.0E
*Kunlun Shankou	PASS	C	35.6N	94.1E
Kuodaba	PPL	C	32.5N	104.3E

name	designation	area	lat.	long.
*Kuqa	PPL	C	41.7N	82.9E
Kuqa Chang	PPL	C	41.1N	83.3E
Kuqa He	RIV	C	42.2N	83.0E
Kuqqa	PPL	U	30.8N	101.2E
Kuragaty	PPL	U	43.1N	73.0E
Kuraminskiy Khrebet	MTS	U	40.8N	70.2E
Kurday	PPL	U	43.4N	75.0E
Kurgantepa	PPL	U	40.8N	72.8E
Kurkat	PPL	U	40.1N	69.3E
Kurkung	PPL	U	33.0N	95.0E
Kurqên	PPL	C	31.8N	83.8E
Kurshab	RIV	U	40.4N	73.3E
Kuruk Gol	RIV	C	42.6N	92.8E
Kuruksay	PPL	U	40.7N	69.3E
Kuruk-Say, Gory see: Gory Kuruk-Say				
*Kuruktag	MTS	C	41.5N	88.5E
Kurumduk	RIV	U	40.5N	76.4E
Kurumdy, Gora see: Gora Kurumdy				
Kurusay see: Kuruksay				
Kusma	PPL	N	28.2N	83.7E
Kuturgu	PPL	U	42.8N	78.1E
Kuva	PPL	U	40.5N	72.1E
Kuvasay	PPL	U	40.3N	72.0E
Kuybyshevo see: Rishtan				
Kuyganyar	PPL	U	40.8N	72.3E
Kuylyu see: Këolyuu				
Kuylyutau, Khrebet see: Khrebet Kerlyuu-Too				
Küysu	PPL	C	43.5N	93.4E
*Küzlak	PPL	C	40.6N	87.4E
Kyamronggag	PPL	C	32.5N	96.3E
Kyangba	PPL	C	30.4N	100.3E
*Kyangngoin	PPL	C	31.2N	95.9E
Kyêbxang Co	LK	C	32.4N	90.0E
*Kyelang	PPL	I	32.6N	77.1E
Kyêrdo	PPL	C	29.5N	86.2E

name	designation	area	lat.	long.
Kyêwu	PPL	C	32.5N	98.3E
Kyikug	PPL	C	36.1N	100.7E
Kyiling	PPL	C	35.2N	100.1E
Kyinzhi	PPL	C	34.1N	100.1E
Kyirong see: Gyirong				
Kyude	PPL	C	31.5N	102.8E
Kyumyush-Tak	RIV	U	42.4N	71.8E
Kyumyush-Tak, Pik see: Pik Kyumyush-Tak				
Kyungëy-Ala-Too, Khrebet see: Khrebet Kyungëy-Ala-Too				
Kyunglung	PPL	C	31.0N	80.5E
Kyurmentyu, Pereval see: Pereval Kyurmentyu				
Kyztogansay	RIV	U	43.0N	73.2E
Kyzyl-Art, Pereval see: Pereval Kyzyl-Art				
Kyzyldala	PPL	U	42.2N	69.4E
Kyzyldangi, Gora see: Gora Kyzyldangi				
Kyzyl-Dzhar	PPL	U	41.3N	72.0E
*Kyzyl-Kiya	PPL	U	40.4N	72.1E
*Kyzyl-Oy	PPL	U	41.9N	74.2E
Kyzylsharyk	PPL	U	43.6N	78.5E
Kyzylsu see: Kyzyl-Suu				
Kyzyl-Suu	RIV	U	39.5N	72.5E
Kyzyltu see: Kyzyl-Tuu				
*Kyzyl-Tuu	PPL	U	42.2N	76.7E
Lab	PPL	C	33.2N	97.1E
*Lablung La	PASS	C	29.6N	84.3E
*Labrang	⊙PPL	C	35.2N	102.4E
*Ladakh Range	MTS	I	34.0N	78.0E
*Ladoi	PPL	C	29.4N	86.6E
Ladoiqangma	PPL	C	32.2N	86.0E
Lagkor Co	LK	C	32.0N	84.0E
Lähe	PPL	B	26.3N	95.4E
*Lahore	PPL	P	31.6N	74.3E
Laiyor	PPL	C	29.9N	84.1E
Laji Shan	MTS	C	36.2N	102.0E
*Laji Shan	PK	C	36.3N	101.6E

name	designation	area	lat.	long.
Laka	PPL	C	29.6N	88.6E
*Lakhimpur	PPL	I	28.0N	80.8E
*Lālsot	PPL	I	26.5N	76.4E
Lamadi	PPL	C	27.2N	98.9E
Lamado	PPL	C	31.0N	93.5E
Lamda see: Ngamda				
Lamdo	PPL	C	32.3N	98.9E
Lamjin	PPL	C	42.8N	89.8E
Lampug	PPL	C	27.8N	92.0E
Lanba	PPL	C	27.9N	102.5E
*Lancang Jiang	RIV	C	30.2N	97.8E
*La'nga Co	LK	C	30.7N	81.2E
Langar	PPL	A	36.7N	72.0E
Langar	PPL	A	37.0N	73.8E
Langgar	PPL	C	29.8N	93.6E
Langjiang Co	LK	C	28.7N	85.8E
Langju	PPL	C	32.3N	80.3E
Langma	PPL	C	30.0N	87.5E
Langmai	PPL	C	30.7N	98.8E
*Langmar	PPL	C	32.3N	79.8E
Langmusi see: Dagcanglhamo				
Langqên Kanbab see: Langqên Zangbo				
*Langqên Zangbo	RIV	C	31.5N	79.5E
Langru	PPL	C	36.8N	79.6E
Längtäng Himäl	MTS	N	28.2N	85.5E
Langxinshan	PPL	C	41.0N	100.3E
Langgazhoinkang	PPL	C	30.6N	81.3E
*Lanping Xian	CCN	C	26.4N	99.2E
*Lanzhou	PPL	C	36.0N	103.7E
Laocheng	PPL	C	25.6N	101.9E
Laojie	PPL	C	25.7N	99.2E
*Laojun Shan	PK	C	26.6N	99.9E
Laoqitai	PPL	C	43.8N	89.9E
Laowo	PPL	C	25.8N	99.0E
Laoximiao	PPL	C	41.5N	100.5E

44

name	designation	area	lat.	long.
Machi Hu see: Ulan Ul Hu				
Madeng	PPL	C	26.4N	99.5E
Mādhubani	PPL	I	26.3N	86.1E
Madian	PPL	C	25.9N	102.9E
*Madoi	PPL	C	35.0N	96.3E
Madoi Xian	CCN	C	34.8N	98.2E
Mādrasah	PPL	A	37.1N	71.1E
Ma'erbang	PPL	C	31.2N	101.9E
Magh Nawul	PPL	A	36.1N	71.0E
*Magitang	⊙PPL	C	35.9N	102.0E
*Mahābhārat Lekh	MTS	N	27.7N	85.4E
Māhai	PPL	C	38.0N	94.4E
Mahai Nongchang	PPL	C	38.0N	94.5E
*Mahākāli	RIV	N	29.5N	80.3E
Mahe	PPL	C	35.3N	104.6E
*Mahendranagar	PPL	N	29.0N	80.2E
Mahezhen see: Mahe				
Mahuanggou	PPL	C	37.3N	95.6E
Maijiaji	PPL	C	35.3N	103.2E
*Mailāni	PPL	I	28.3N	80.3E
Māima	PPL	C	33.6N	102.0E
Mainda	PPL	C	32.0N	97.2E
*Maindong	⊙PPL	C	31.0N	85.1E
Maindu	PPL	C	31.5N	91.6E
Maingkwan	PPL	B	26.3N	96.6E
Mainkung	PPL	C	28.6N	98.2E
Mainling Xian	CCN	C	29.2N	94.1E
Mainpu	PPL	C	28.7N	86.2E
*Mainpuri	PPL	I	27.2N	79.0E
Maiqu Zangbo	RIV	C	29.7N	87.6E
Maiyü	PPL	C	30.1N	97.4E
Maizha	PPL	C	32.3N	103.2E
Maizhokunggar Xian	CCN	C	29.8N	91.7E
Maji	PPL	C	27.3N	98.8E
Majiahewan	PPL	C	37.1N	105.8E

name	designation	area	lat.	long.
Majiaji	PPL	C	35.3N	103.6E
Majiayao	PPL	C	35.3N	103.3E
*Majie	PPL	C	25.7N	102.9E
*Makalu	PK	X	27.9N	87.1E
Makaru Shan see: Makalu				
Makou	PPL	C	27.6N	103.3E
Ma La	PASS	A	28.5N	90.1E
*Ma La	PASS	C	28.9N	85.3E
*Malakand	PPL	P	34.6N	71.9E
Malakwāl	PPL	C	32.6N	73.2E
Malangwa	PPL	N	26.8N	85.6E
Malaut	PPL	I	30.2N	74.5E
Malayiwan	PPL	C	34.9N	98.4E
*Mali	RIV	B	26.7N	97.8E
Mali	PPL	C	27.5N	102.2E
Maliang	PPL	C	43.9N	91.5E
Malianjing	PPL	C	38.7N	101.3E
*Malianjing	PPL	C	41.5N	95.2E
Malianquan	PPL	C	38.5N	102.6E
*Maliliang	PPL	C	31.9N	89.1E
Malong	PPL	C	26.9N	101.8E
Malong Feng	PK	X	25.5N	100.0E
Malong Xian	CCN	C	25.4N	103.6E
Malovodnoye	PPL	U	43.5N	77.7E
Mālpura	PPL	I	26.3N	75.4E
Malutang	PPL	C	26.1N	102.5E
Malyy Naryn	RIV	U	41.5N	76.5E
*Mamat Kuduk	SPNG	C	40.3N	77.5E
Mamba	PPL	C	30.1N	92.2E
Maming see: Mamingzhen				
Mamingzhen	PPL	C	30.1N	103.3E
Mamta	PPL	C	32.4N	94.3E
Manāli	PPL	I	32.2N	77.2E
Manang	PPL	N	28.7N	84.0E
Māna Pass	PASS	X	31.1N	79.5E

name	designation	area	lat.	long.
Manasarowar Lake *see:* Mapam Yumco				
Manas He	RIV	C	43.5N	85.5E
*Manaslu	PK	N	28.6N	84.6E
Mandalik	PPL	C	38.5N	89.7E
*Mandi	PPL	I	31.7N	76.9E
Mandi Būrewāla	PPL	P	30.2N	72.7E
Mandi Dabwāli	PPL	I	29.9N	74.7E
*Mandra	PPL	P	33.4N	73.2E
Mangaldai	PPL	I	26.5N	92.0E
Mangkuan	PPL	C	25.4N	98.8E
Mangkyi	PPL	B	26.3N	98.3E
Mangnai	PPL	C	37.8N	91.7E
Mangnai Zhen	PPL	C	38.3N	90.1E
Mangra	PPL	C	29.6N	89.6E
*Mangra	⊙PPL	C	35.5N	100.7E
Mangra Qu	RIV	C	35.6N	100.7E
Mani	PPL	C	34.8N	87.2E
*Maniganggo	PPL	C	31.9N	99.2E
Manin Qagan Tungge	PPL	C	40.5N	101.3E
Mankent *see:* Manket				
Manket	PPL	U	42.4N	69.8E
Mānsehra	PPL	P	34.4N	73.2E
Maobosheng	PPL	C	37.0N	101.5E
Maoduo	PPL	C	35.3N	100.1E
Mao'ergai	PPL	C	32.6N	103.0E
Mao'ergai He	RIV	C	32.4N	103.1E
Maojie	PPL	C	25.4N	101.6E
Maojie	PPL	C	25.5N	102.1E
*Maomao Shan	PK	C	37.1N	103.1E
Maoniugou	PPL	C	30.6N	101.7E
Maoniupo	PPL	C	34.3N	86.3E
Maoniu Shan	MT	C	33.1N	104.0E
Maomiushan	PPL	C	36.8N	97.8E
Maowen Qiangzu Zizhixian	CCN	C	31.6N	103.8E
Maoxiangba	PPL	C	32.5N	104.7E

name	designation	area	lat.	long.
Maozangsi	PPL	C	37.5N	102.4E
Maozu	PPL	C	27.4N	102.9E
*Mapam Yumco	LK	C	30.6N	81.4E
Mapisi	PPL	C	37.0N	100.8E
Maqên	PPL	C	31.6N	90.3E
*Maqên Gangri	PK	C	34.8N	99.4E
Maqên Xian	CCN	C	34.5N	100.2E
*Ma Qu	RIV	C	33.8N	101.2E
Maquan He	RIV	C	29.8N	83.6E
Maqu Xian	CCN	C	34.0N	102.1E
Maralwexi *see:* Bachu				
*Mardān	PPL	P	34.2N	72.0E
Mardêng	PPL	C	32.3N	91.2E
Marê	PPL	C	30.7N	100.1E
Margai Caka	LK	C	35.1N	86.7E
Margam Ri	MTS	C	33.2N	85.7E
*Margilan	PPL	U	40.5N	71.7E
*Margog Caka	LK	C	33.8N	86.9E
*Maryang	PPL	C	29.7N	89.9E
Markam Shan	MTS	C	29.9N	98.5E
Markam Xian	CCN	C	29.6N	98.5E
Markansu	RIV	U	39.3N	73.5E
Markhamat	PPL	U	40.5N	72.3E
*Markit	PPL	C	38.9N	77.6E
Markog	PPL	C	33.1N	100.4E
*Markog He	RIV	C	32.5N	101.5E
Marmê	PPL	C	32.0N	83.6E
Marqu	PPL	C	29.8N	90.7E
Mar Qu *see:* Markog He				
*Marri	PPL	C	30.9N	96.3E
Maru	PPL	C	34.6N	102.1E
*Mārwār	PPL	I	25.7N	73.7E
Maryang	PPL	C	37.3N	75.7E
*Marzhing	PPL	C	33.3N	100.4E
Masak	PPL	U	43.6N	78.3E

49

name	designation	area	lat.	long.
Masanchi	PPL	U	42.9N	75.3E
Mashutang	PPL	C	26.7N	103.2E
*Mastūj	PPL	P	36.3N	72.5E
Matajing	PPL	C	29.5N	103.9E
Matang	PPL	C	31.9N	102.6E
Matcha	PPL	U	40.4N	69.3E
*Mathura	PPL	I	27.5N	77.7E
Matisi	PPL	C	38.4N	100.4E
Matizi	PPL	C	33.2N	103.4E
Mawu	PPL	C	34.4N	104.9E
Maxia	PPL	C	32.4N	98.1E
*Maxian Shan	PK		35.7N	103.9E
Maxian Shan	MTS	C	35.8N	103.8E
Maxiapkol	PPL	C	41.4N	85.2E
Maya	PPL	C	33.9N	103.7E
Mayakovskogo, Pik see: Pik Mayakovskogo				
Maydantal	RIV	U	42.2N	70.7E
*Mayêr Kangri	PK	C	33.4N	86.8E
Maying	PPL	C	35.3N	105.0E
Maying	PPL	C	36.0N	102.8E
Maying	PPL	C	38.3N	101.1E
Maying	PPL	C	39.3N	99.2E
Mayingsi	PPL	C	36.6N	102.6E
Mayli-Say	PPL	U	41.3N	72.4E
Mayly-Suu	RIV	U	41.3N	72.5E
Maymak	PPL	U	42.7N	71.2E
Mäymey	PPL	A	38.4N	71.0E
*Mazar	PPL	C	36.4N	77.2E
*Mazartag	PPL	C	38.4N	80.8E
Mazartag	PK	C	38.6N	80.1E
Mazhanjie	PPL	C	25.2N	98.4E
Mazichuan	PPL	C	34.2N	104.1E
*Mazong Shan	PK	C	41.5N	97.1E
Mazong Shan	MTS	C	41.7N	97.0E
Medenshor	PPL	U	37.3N	71.8E

name	designation	area	lat.	long.
*Mêdog	PPL	C	29.2N	95.3E
*Mêdog	PPL	C	30.8N	85.1E
Mêdogdêng	PPL	C	30.0N	96.0E
*Meerut	PPL	I	29.0N	77.7E
Mê'gyai	PPL	C	35.8N	101.6E
Meichuan	PPL	C	34.5N	104.0E
Meigu He	RIV	C	28.1N	103.2E
*Meigu Xian	CCN	C	28.3N	103.1E
*Meishan	PPL	C	30.0N	103.9E
Meiwu	PPL	C	35.0N	103.0E
*Meixing	⊙PPL	C	30.9N	102.3E
Meiyu	PPL	C	27.4N	101.3E
*Mêmar Co	LK	C	34.2N	82.2E
Mêmo	PPL	C	28.5N	86.8E
Mendong Gompa see: Maindong				
Menggubao	PPL	C	36.6N	78.1E
Menghugang	PPL	C	29.3N	102.2E
Mengjiaqiao	PPL	C	40.0N	94.6E
Mengxuan	PPL	C	35.8N	105.5E
Mengyangzhen	PPL	C	30.9N	104.1E
Menyuan Huizu Zizhixian	CCN	C	37.4N	101.6E
Mêqu	PPL	C	33.9N	102.3E
Mê Qu see: Hei He				
Meridianal'nyy Khrebet	MTS	X	42.3N	80.3E
Merke	PPL	U	42.9N	73.2E
Merke	RIV	U	42.7N	73.3E
Mêrqung Co	LK	C	31.0N	84.6E
Merta	PPL	I	26.7N	74.1E
*Mêruma	PPL	C	32.8N	101.9E
Mêwa	PPL	C	33.0N	102.8E
Meymik	PPL	A	38.2N	70.6E
Mezhdurechenskoye	PPL	U	43.5N	76.8E
Miancaowan	PPL	C	35.2N	98.9E
*Mianmian Shan	MTS	C	27.3N	100.8E
*Mianning	PPL	C	28.5N	102.1E

Mosuoying	PPL	C	27.0N	102.2E
Motian Ling	MTS	C	32.8N	104.6E
*Motihāri	PPL	I	26.7N	84.9E
*Mount Everest	PK	X	27.9N	86.9E
Mount Godwin Austen see: K2				
Moxi	PPL	C	29.6N	102.1E
Moximian	PPL	C	29.6N	102.1E
*Moyu	PPL	C	37.3N	79.7E
Mozhong	PPL	C	32.3N	96.8E
Mucheng	PPL	C	29.8N	103.4E
Mucheng	PPL	C	31.6N	102.3E
*Muchuan	PPL	C	28.9N	104.1E
Mudan	PPL	C	34.4N	105.5E
Muganhe	PPL	C	28.1N	103.9E
Mugarripug	PPL	C	32.3N	87.5E
Muggar Kangri	MTS	C	32.3N	87.7E
Mug Qu	RIV	C	33.8N	93.7E
Muguaping	PPL	C	31.3N	104.0E
Mugxung	PPL	C	33.2N	94.3E
Muhar	PPL	C	36.8N	98.9E
Muhor see: Qoijê				
Muhur	PPL	C	43.2N	82.6E
Muji	PPL	C	37.4N	78.5E
Muji	PPL	C	39.0N	74.4E
Mükangsar	PPL	C	29.5N	87.6E
Muksu	RIV	U	39.2N	72.0E
Mula	PPL	C	29.6N	100.6E
Muli Zangzu Zizhixian	CCN	C	27.8N	101.3E
*Multān	PPL	P	30.2N	71.5E
Muminabad see: Leningradskiy				
Mun	PPL	U	37.6N	71.7E
Mun'ke	RIV	U	43.1N	73.0E
*Murgab	PPL	U	38.2N	73.9E
*Murgab	RIV	U	38.3N	73.5E
Muri	PPL	C	37.6N	100.7E

*Muri	PPL	C	38.1N	99.0E
Murree	PPL	P	33.9N	73.4E
Muzaffarābād	PPL	P	34.4N	73.5E
*Muzaffarnagar	PPL	I	29.5N	77.7E
*Muzat He	RIV	C	41.7N	81.3E
Muzkol	RIV	U	38.8N	73.5E
Muzkol, Khrebet see: Khrebet Muzkol				
Muzluk	PPL	C	37.1N	87.6E
*Muztag	PK	C	36.1N	80.3E
Muztagata	PK	C	38.2N	75.1E
*Muztag Feng	PK	C	36.4N	87.3E
Muztag Shan	MTS	C	36.6N	87.6E
Myrza-Ake	PPL	U	40.8N	73.4E
Myudyuryum	RIV	U	41.0N	76.9E
Nabru see: Namru				
Nabu	PPL	C	29.1N	101.7E
Nabzain	PPL	C	38.5N	98.0E
Nadong	PPL	C	28.5N	101.4E
*Nagarzê	PPL	C	29.0N	90.4E
*Nāgaur	PPL	I	27.2N	73.8E
Nagchhukha see: Nagqu				
Nagchu Dzong see: Nagqu				
Nagina	PPL	I	29.4N	78.5E
Nagir	PPL	P	36.3N	74.7E
Nagiog	PPL	C	30.5N	96.1E
Nagmung	PPL	B	27.5N	97.8E
Nagpag Co	LK	C	27.8N	99.6E
Nag Qu	RIV	C	31.2N	92.1E
*Nagqu	PPL	C	31.5N	92.1E
Nagza	PPL	C	38.6N	97.0E
Nahan	PPL	I	30.6N	77.3E
Naij Gol He	RIV	U	35.8N	94.0E
Naij Tal	PPL	C	35.9N	94.5E
*Nailung	RIV	U	28.6N	86.4E
Nailung	PPL	C	29.1N	93.8E

name	designation	area	lat.	long.
*Naini Tāl	PPL	I	29.4N	79.4E
Naiqu	PPL	C	28.7N	101.6E
Nairi	PPL	C	27.8N	99.6E
Naixi	PPL	C	28.3N	91.3E
Naixi	PPL	C	29.7N	94.2E
Naiyü	PPL	C	29.1N	94.1E
Naizha	PPL	C	31.3N	96.5E
*Najībābād	PPL	I	29.6N	78.3E
Nakchu see: Biru				
Nakshö see: Biru				
Nala'nga	PPL	C	31.2N	90.3E
Naluzongba	PPL	C	29.3N	100.7E
*Namangan	PPL	U	41.0N	71.7E
*Nämche	PPL	N	27.8N	86.7E
*Nam Co	LK	C	30.8N	90.8E
Namco	PPL	C	30.8N	91.1E
Namda	PPL	C	32.4N	101.0E
Namgyaigang	PPL	C	29.5N	88.5E
Namgyaixoi	PPL	C	29.1N	91.1E
*Namjagbarwa Feng	PK	C	29.6N	95.0E
Nāmjā La see: Namsê La				
*Namka	⊙PPL	C	29.6N	90.9E
*Namling	PPL	C	29.6N	89.0E
Namoding	PPL	C	30.3N	86.8E
Namoqê	PPL	C	29.5N	88.3E
*Namru	PPL	C	31.9N	80.1E
Namsai	PPL	C	32.8N	95.7E
Namsê La	PASS	X	29.9N	82.5E
Namsê Shankou see: Namsê La				
Namtang	PPL	C	35.2N	99.6E
Nanba	PPL	C	32.1N	104.7E
Nanbaxian	PPL	C	37.9N	94.3E
*Nanda Devi	PK	I	30.4N	80.0E
*Nang	PPL	C	29.0N	93.0E
*Nanga Parbat	PK	P	35.3N	74.6E

name	designation	area	lat.	long.
Nangdoi	PPL	C	36.1N	102.0E
Nangkartse Dzong see: Nagarzê				
Nangqên Xian	CCN	C	32.2N	96.4E
Nangucheng	PPL	C	38.5N	100.5E
Nanhu	PPL	C	39.9N	94.1E
Nanhua	PPL	C	39.3N	99.7E
*Nan Hulsan Hu	LK	C	36.7N	95.8E
Nânpâra	PPL	I	27.8N	81.5E
*Nanping	PPL	C	33.2N	104.2E
Nanquan	PPL	C	40.9N	98.4E
Nanrgang	PPL	C	28.9N	102.6E
Nansena, Pik see: Pik Nansena				
Nanshankou	PPL	C	43.2N	93.7E
Nantan	PPL	C	37.2N	101.4E
*Nanyun	PPL	B	26.9N	96.2E
Naomaohu	PPL	C	43.7N	94.9E
Naoshera	PPL	I	33.2N	74.3E
*Napug	⊙PPL	C	32.4N	81.1E
Naqên	PPL	C	31.9N	84.0E
Naran Bulag	PPL	C	41.4N	106.4E
*Narang	PPL	A	34.8N	71.1E
*Naran Sebstein Bulag	SPNG	C	42.7N	96.9E
Narat	PPL	C	43.3N	84.0E
Narat Shan	MTS	C	43.1N	83.4E
Nāray	PPL	A	35.2N	71.5E
Nārāyani	RIV	N	27.7N	84.3E
Narendranagar	PPL	I	30.2N	78.4E
Narimanov	PPL	U	41.2N	69.2E
Narin Go	PPL	C	38.9N	96.5E
Narin Gol	RIV	C	37.0N	93.0E
Narin He see: Dong He				
*Nārnaul	PPL	I	28.1N	76.2E
Nārowāl	PPL	P	32.1N	74.9E
Narüdo	PPL	C	31.4N	92.9E
*Naryn	RIV	U	41.4N	74.5E

name	designation	area	lat.	long.
Nyima	PPL	C	31.5N	92.7E
Nyima	PPL	C	31.7N	92.2E
*Nyima	PPL	C	31.9N	87.8E
Nyimda	PPL	C	32.6N	101.3E
Nyinba	PPL	C	31.7N	93.7E
Nyinba	PPL	C	34.4N	103.0E
Nyingchi Xian	CCN	C	29.5N	94.5E
Nyingda	PPL	C	29.8N	93.4E
*Nyingzhong	PPL	C	30.3N	90.8E
Nyinqug	PPL	C	35.2N	100.8E
Nyixung	PPL	C	30.7N	85.7E
Nyizhê	PPL	C	31.0N	89.0E
Nyogzê	PPL	C	30.7N	82.1E
Nyugku	PPL	C	29.5N	84.6E
Obikhingou	RIV	U	38.8N	70.9E
Obimazar	RIV	U	38.4N	70.0E
Obimazar	RIV	U	38.7N	71.4E
*Obo	PPL	C	37.9N	100.9E
*Obo Liang	PPL	C	38.8N	92.7E
Obo Nanshan	MTS	C	37.8N	100.9E
Obruchëvka	PPL	U	42.5N	69.1E
Obson Jah	PPL	C	37.7N	94.2E
Oglung He	RIV	C	35.2N	86.1E
Ohsalur	PPL	C	39.6N	74.8E
Oibab	PPL	C	29.5N	97.5E
Oiga	PPL	C	29.2N	92.2E
Oilê	PPL	C	29.3N	101.4E
Oi Qu see: Yu Qu				
Oisêr	PPL	C	33.6N	99.8E
Oiyug	PPL	C	29.6N	89.6E
*Okāra	PPL	P	30.8N	73.5E
*Okhaldhunga	PPL	N	27.3N	86.5E
Oksoy, Gora see: Gora Oksoy				
Oksu	RIV	U	38.2N	74.2E
Oktyabr'skoye	PPL	U	41.1N	73.1E

name	designation	area	lat.	long.
Oku	PPL	C	36.3N	80.3E
Olon Bulag	PPL	C	40.5N	106.3E
*Oma	PPL	C	32.3N	83.2E
Ombu	PPL	C	31.3N	86.7E
Omxa	PPL	C	36.1N	79.8E
On-Archa	RIV	U	41.6N	76.0E
Ondor Mod	PPL	C	41.1N	104.3E
*Ongt Gol	PPL	C	41.4N	102.1E
Opal	PPL	C	39.2N	75.5E
*Orai	PPL	I	26.0N	79.4E
*Orba Co	LK	C	34.5N	81.0E
Orlovka	PPL	U	42.8N	75.6E
Orma	PPL	C	28.8N	85.2E
Orto-Nura	PPL	U	41.4N	76.2E
*Orto-Tokoy	PPL	U	42.3N	76.0E
Orto-Tokoyskoye Vodokhranilishche	RSV	U	42.4N	75.9E
*Osh	PPL	U	40.5N	72.8E
Oshnügän	VAL	A	36.3N	70.7E
Osiyän	PPL	I	26.8N	72.9E
Otmëk, Pereval see: Pereval Otmëk				
Ottuk	PPL	U	42.3N	76.3E
Oygaing	RIV	U	42.2N	70.8E
Oyrandy	RIV	U	43.0N	73.6E
Oy-Tal	PPL	U	40.4N	74.1E
Oy-Tal	RIV	U	40.5N	74.0E
*Oytal	PPL	U	42.9N	73.3E
Oytar	PPL	C	38.9N	75.4E
Oytograk	PPL	C	36.8N	81.9E
*Ozero Chatyr-Kël'	LK	U	40.7N	75.3E
*Ozero Issyk-Kul'	LK	U	42.4N	77.3E
*Ozero Karakul'	LK	U	39.0N	73.5E
Ozero Rangkul'	LK	U	38.5N	74.3E
Ozero Sarezskoye	LK	U	38.2N	72.8E
Ozero Shorkul'	LK	U	38.5N	74.2E

name	designation	area	lat.	long.
*Ozero Song-Kël'	LK	U	41.8N	75.2E
Ozero Sonkël' see: Ozero Song-Kël'				
Ozero Turumtaykul'	LK	U	37.5N	72.5E
Ozero Tuzkol'	LK	U	43.0N	80.0E
*Ozero Yashil'kul'	LK	U	37.8N	72.8E
Ozero Zorkul'	LK	X	37.4N	73.7E
Ozgëryush	PPL	U	41.1N	74.6E
Ozgeryush	PPL	U	42.6N	72.6E
Ozgorush see: Ozgëryush				
Ozgorush see: Ozgëryush				
Paca	PPL	C	30.5N	88.7E
Padain	PPL	C	31.4N	95.0E
*Padam	PPL	I	33.5N	76.9E
Padysha-Ata	RIV	U	41.5N	71.5E
Pagbar	PPL	C	31.8N	101.9E
Paggai	PPL	C	30.3N	94.3E
Paggarama see: Zhasa				
Paggor	⊙PPL	C	32.8N	91.3E
*Pagnag	⊙PPL	C	29.5N	90.0E
*Pagqên	PPL	C	32.2N	91.6E
*Pagri	PPL	C	27.7N	89.1E
Pahalgâm	PPL	I	34.0N	75.4E
*Paikü Co	LK	C	28.9N	85.5E
Paikü Kangri	MTS	C	28.5N	85.5E
Pailou	PPL	C	36.6N	103.0E
Painbo Nongchang	PPL	C	29.8N	91.2E
Paingar	PPL	C	31.2N	94.0E
*Paiñqi	PPL	C	29.7N	83.5E
Paiqu	PPL	U	29.5N	94.8E
Pakhtaabad	PPL	U	40.9N	72.5E
Pâkpattan	PPL	P	30.4N	73.4E
Pal see: Bar				
Palas	PPL	U	40.3N	69.8E
Pâli	PPL	I	25.8N	73.4E
*Palung Co	LK	C	30.9N	83.5E

name	designation	area	lat.	long.
Palvantash	PPL	U	40.6N	72.2E
Pamir	RIV	X	37.2N	73.2E
*Pamir	MTS	X	38.3N	73.6E
Panâm	PPL	A	35.9N	70.9E
Pan'anzhen	PPL	C	34.7N	105.1E
*Pangong Tso	LK	I	33.7N	78.7E
*Pangsâu Pass	PASS	X	27.3N	96.2E
Pânipat	PPL	I	29.4N	77.0E
Panj, Daryâ-ye see: Daryâ-ye Panj				
Panpo	PPL	C	37.6N	101.3E
Paomaping	PPL	C	26.9N	100.9E
Pap	PPL	U	40.9N	71.1E
Paprowk	PPL	A	35.6N	71.2E
Par	PPL	C	32.6N	92.3E
Parâsi	PPL	N	27.5N	83.7E
Parco	PPL	C	32.3N	82.1E
Parding	PPL	C	32.8N	88.6E
Parling	PPL	C	30.3N	90.8E
Parlung Zangbo	RIV	C	30.0N	95.0E
*Parma	PPL	C	31.2N	84.2E
Parma	PPL	C	34.0N	101.9E
Parngain	PPL	C	30.7N	102.0E
Paro	PPL	Z	27.4N	89.5E
*Parta	PPL	C	31.5N	91.4E
Parxixingka	PPL	C	36.4N	99.2E
Paryang	PPL	C	30.1N	83.4E
Pâsighât	PPL	I	28.1N	95.3E
Pastkhuf	PPL	U	37.8N	71.6E
*Pasu	PPL	P	36.5N	74.9E
Pâtan	PPL	N	27.6N	85.3E
Pathâlipâm	PPL	I	27.5N	94.4E
Pathânkot	PPL	I	32.3N	75.7E
*Patiâla	PPL	I	30.3N	76.5E
Patkaglik see: Patkaklik				
*Pâtkai Range	MTS	X	26.7N	95.6E

name	designation	area	lat.	long.
Patkaklik	PPL	C	37.3N	87.0E
Patkhor, Pik see: Pik Patkhur				
Patkhur, Pik see: Pik Patkhur				
Pauri	PPL	I	30.1N	78.8E
Paytug	PPL	U	40.9N	72.2E
Payzawat see: Jiashi				
*Pazhug	PPL	C	28.4N	87.0E
Pêdo La	PASS	X	29.2N	83.4E
Pêdo Shankou see: Pêdo La				
Pei	PPL	C	29.5N	94.8E
Peijiaying	PPL	C	37.4N	103.5E
*Pêlung	PPL	C	30.1N	95.1E
Pencuodarjihelang	PPL	C	36.0N	98.1E
Pênda	PPL	C	32.6N	97.5E
Pengshan	PPL	C	30.2N	103.8E
Peng Xian	CCN	C	30.9N	103.9E
Pengzhen	PPL	C	30.5N	103.8E
Pereval Akbaytal	PASS	U	38.6N	73.6E
Pereval Ala-Bel'	PASS	U	42.3N	73.1E
Pereval Almaaty	PASS	U	43.0N	76.8E
Pereval Almaty see: Pereval Almaaty				
Pereval Bedel'	PASS	U	41.4N	78.4E
Pereval Chaty-Bel'	PASS	U	41.0N	78.2E
Pereval Chiyirchik see: Pereval Chyyyrchyk				
Pereval Chyyyrchyk	PASS	U	40.3N	73.5E
Pereval Dolon	PASS	U	41.8N	75.8E
Pereval Khodzhent	PASS	X	40.3N	76.2E
Pereval Kogart	PASS	X	40.3N	74.8E
Pereval Koytezek	PASS	U	37.5N	72.8E
Pereval Kyurmentyu	PASS	U	42.9N	78.2E
Pereval Kyzyl-Art	PASS	U	39.4N	73.3E
Pereval Nayzatash	PASS	U	37.9N	73.8E
Pereval Otmёk	PASS	U	42.3N	73.2E
Pereval Taldyk	PASS	U	39.8N	73.2E
Pereval Torugart see: Turugart Shankou				
Pereval Turugart see: Turugart Shankou				
Pereval Tyuzashu see: Pereval Tyuz-Ashuu				
Pereval Tyuz-Ashuu	PASS	U	42.3N	73.3E
Pereval Uz-Bel' see: Uzbel Shankou				
Pervomayevka	PPL	U	42.1N	69.9E
Pervomayskiy see: Ak-Suu			42.8N	74.1E
*Peshäwar	PPL	P	34.0N	71.6E
Petra Pervogo, Khrebet see: Khrebet Petra Pervogo				
Pexung	PPL	C	32.2N	92.6E
*Phalodi	PPL	I	27.1N	72.4E
Pharenda	PPL	I	27.1N	83.3E
*Phidim	PPL	N	27.2N	87.8E
Phuntsholing	PPL	Z	26.9N	89.4E
Pianjiao	PPL	C	26.0N	100.6E
Pianjiaojie see: Pianjiao				
Pianma	PPL	C	26.0N	98.5E
Pich, Darrah-ye see: Darrah-ye Pich				
Pik Bazarteppa	PK	U	38.0N	73.1E
Pik Dankova	PK	X	41.0N	77.6E
Pik Frunze	PK	U	39.0N	72.8E
Pik Garmo	PK	U	38.8N	72.1E
Pik Igla	PK	U	39.5N	70.6E
Pik Karasak	PK	U	38.8N	73.6E
*Pik Karla Marksa	PK	U	37.2N	72.4E
Pik Khan-Tengri	PK	X	42.2N	80.2E
*Pik Kommunizma	PK	U	38.9N	72.0E
Pik Kumyshtag see: Pik Kyumyush-Tak				
Pik Kyumyush-Tak	PK	U	42.3N	71.8E
*Pik Lenina	PK	U	39.3N	72.9E
Pik Mayakovskogo	PK	U	37.0N	71.7E
Pik Moskva	PK	U	38.9N	71.9E
Pik Nansena	PK	U	42.2N	79.6E
Pik Patkhor see: Pik Patkhur				
Pik Patkhur	PK	U	37.9N	72.2E
Pik Piramidal'nyy	PK	U	39.5N	70.2E

59

name	designation	area	lat.	long.
Pskent	PPL	U	40.9N	69.3E
*Pubu	⊙PPL	C	31.3N	90.0E
*Pūdog Zangbo	RIV	C	30.4N	84.6E
*Pudu He	RIV	C	26.0N	102.7E
*Pūgal	PPL	I	28.5N	72.7E
*Puge Xian	CCN	C	27.4N	102.5E
Pugyülung	PPL	C	31.5N	100.0E
Puhuy	PPL	C	41.3N	85.9E
Pujiang	PPL	C	30.2N	103.5E
*Pula	⊙PPL	C	29.5N	94.5E
Pula	PPL	C	30.0N	89.3E
Puladi	PPL	C	27.6N	98.8E
Pulu	PPL	C	36.2N	81.4E
*Puma Yumco	LK	C	28.6N	90.3E
*Pum Qu	RIV	C	28.6N	87.1E
*Punakha	PPL	Z	27.6N	89.8E
*Pünch	PPL	I	33.8N	74.1E
Püncogling	PPL	C	29.3N	88.0E
Pungan	PPL	U	40.7N	70.8E
Pung Co	LK	C	31.5N	90.9E
Punzom	PPL	C	31.0N	93.7E
Pupeng	PPL	C	25.3N	100.9E
Pupiao	PPL	C	25.0N	99.0E
Puqu	PPL	C	29.4N	94.3E
Purang see: Burang				
Pur Co	LK	C	34.9N	81.9E
Purgadala	PPL	C	33.6N	84.3E
*Purog Kangri	PK	C	33.9N	89.2E
Pusa	PPL	B	36.4N	79.0E
*Putao	PPL	B	27.4N	97.4E
Puwei	PPL	C	27.0N	101.9E
Puxiong	PPL	C	28.5N	102.6E
Puyang	PPL	C	31.0N	103.7E
Puzang	PPL	C	29.0N	96.5E
*Pyandzh	RIV	U	38.2N	70.4E

name	designation	area	lat.	long.
*Qabgar	PPL	C	29.7N	93.9E
Qabnag	PPL	C	29.4N	94.4E
*Qabqa	⊙PPL	C	36.2N	100.6E
Qagan Bulag	PPL	C	38.3N	104.6E
Qagan Ders	PPL	C	41.5N	106.3E
Qagan Gu	PPL	C	38.7N	96.4E
*Qagan Nur	PPL	C	36.9N	98.8E
Qagan Nur	PPL	C	43.0N	86.1E
Qagan Obot Ling	MTS	C	38.5N	95.7E
Qagan Qonj	WELL	C	42.1N	96.7E
Qagan Tohoi	PPL	C	35.9N	94.7E
*Qagan Us	⊙PPL	C	36.3N	98.1E
Qagan Us He	RIV	C	36.1N	98.5E
Qagbasêrag	PPL	C	30.8N	92.7E
Qagca	PPL	C	32.7N	98.3E
*Qagcaka	PPL	C	32.6N	81.8E
Qagchêng Xian see: Xiangcheng Xian				
Qagong Co	LK	C	34.5N	82.4E
Qagzê	PPL	C	31.6N	93.1E
Qaidam He	RIV	C	36.4N	96.9E
*Qaidam Pendi	DEPR	C	37.5N	94.4E
Qaidam Shan	MTS	C	37.9N	95.7E
Qaidar see: Cêtar				
Qainaqangma	PPL	C	33.2N	88.7E
*Qajortêbu	⊙PPL	C	30.9N	88.6E
Qakar	PPL	C	36.5N	80.6E
Qal'eh-ye Mirzā Shāh	PPL	A	37.4N	70.9E
Qal'eh-ye Panjeh	PPL	A	37.0N	72.6E
Qalgar	PPL	C	38.0N	104.5E
*Qamalung	PPL	C	34.5N	99.2E
*Qamdo	PPL	C	31.1N	97.2E
Qamdün	PPL	C	30.5N	97.9E
Qammê	PPL	C	33.7N	102.4E
*Qamqên	⊙PPL	C	29.2N	89.7E
Qangba	PPL	C	32.5N	80.7E

name	designation	area	lat.	long.
Shang Mêxoi	PPL	C	31.6N	98.9E
Shang Ngawa	PPL	C	32.9N	101.6E
Shang Paisog	PPL	C	30.2N	101.5E
Shangsuoqiao	PPL	C	31.3N	103.8E
Shangtse see: Qangzê				
Shang Ulastai	PPL	C	43.8N	83.1E
Shangwuzhuang	PPL	C	36.8N	101.4E
Shangyanba	PPL	C	32.1N	104.6E
Shangyou Shuiku	RSV	C	40.4N	80.6E
Shangyou Yichang	PPL	C	40.4N	80.8E
Shangyun	PPL	C	25.0N	98.6E
Shang Zanggor	PPL	C	34.1N	100.2E
*Shang Zayü	PPL	C	28.7N	96.7E
Shangzhou	PPL	C	28.8N	104.0E
Shankou	PPL	C	42.1N	94.1E
Shanlenggang	PPL	C	28.3N	103.4E
*Shanshan	PPL	C	42.8N	90.2E
Shanshanzhan	PPL	C	43.1N	90.4E
Shanyang	PPL	C	25.3N	99.3E
Shaonaomenchehe	PK	C	28.2N	103.2E
Shaoya	PPL	C	36.1N	103.9E
Shaoping	PPL	C	29.9N	103.0E
Shapingguan	PPL	C	31.1N	103.5E
Shaqiao	PPL	C	25.2N	101.1E
Shawan	PPL	C	29.4N	103.5E
Shawan	PPL	C	33.6N	104.5E
Shawanzhen	PPL	C	29.4N	103.8E
Shaxi	PPL	C	26.3N	99.8E
Shaymak	PPL	U	37.6N	74.8E
*Shazaoyuan	PPL	C	39.9N	94.3E
*Shazud	PPL	U	37.7N	72.4E
Sheghnän	PPL	A	37.5N	71.5E
Shejie	PPL	C	25.2N	100.2E
Sheker	PPL	U	42.5N	71.2E
*Shekhüpura	PPL	P	31.7N	74.0E

name	designation	area	lat.	long.
Shela	PPL	C	31.9N	93.3E
Shenduli	PPL	C	34.4N	104.4E
Shengdiwan	PPL	C	40.2N	98.6E
Shengli Daban	PASS	C	43.0N	86.8E
Sheng-li Feng see: Tomur Feng				
Shengli Jiuchang	PPL	C	41.3N	80.6E
Shengilikou	PPL	C	37.5N	95.4E
Shengli Qichang	PPL	C	40.3N	80.0E
Shengli Shibachang	PPL	C	40.6N	81.7E
Shengli Shijiuchang	PPL	C	40.5N	81.5E
Shengli Shiliuchang	PPL	C	40.6N	81.5E
Shengli Shisanchang	PPL	C	40.7N	80.8E
Shenhuguan	PPL	I	25.1N	97.7E
*Sheopur	PPL	I	25.7N	76.7E
Shergarh	PPL	I	26.3N	72.4E
Shetang	PPL	C	34.5N	105.9E
Shibandong Jing see: Shiban Jing				
Shibandun	PPL	C	41.7N	98.5E
*Shiban Jing	WELL	C	40.9N	95.9E
*Shiban Jing	WELL	C	40.8N	94.7E
*Shiban Quan	SPNG	C	40.5N	99.4E
Shibantan	PPL	C	30.7N	104.2E
Shibanxi	PPL	C	29.2N	103.8E
Shibaocheng	PPL	C	39.8N	96.0E
Shibu	PPL	C	34.1N	105.3E
Shichuan	PPL	C	35.1N	104.9E
Shichuan	PPL	C	36.1N	104.0E
Shideng	PPL	C	26.7N	99.1E
*Shidongsi	⊙PPL	C	36.3N	103.9E
Shifang	PPL	C	33.1N	104.5E
Shifang Xian	CCN	C	31.1N	104.1E
Shigatse see: Xigazê				
Shigong	PPL	C	40.4N	95.7E
Shigu	PPL	C	26.8N	99.9E
Shigu	PPL	C	31.5N	103.7E

70

name	designation	area	lat.	long.
Shuimoguan	PPL	C	38.2N	101.8E
Shuiquan	PPL	C	36.8N	104.6E
Shuiquanzi	PPL	C	39.3N	101.6E
Shuixie	PPL	C	25.1N	99.7E
Shuiyazidun	PPL	C	38.7N	93.3E
Shuizhan	PPL	C	38.0N	91.7E
*Shule	PPL	C	39.3N	76.0E
*Shulehe	PPL	C	40.4N	96.8E
*Shule He	RIV	C	40.5N	96.1E
Shule Nanshan	MTS	C	38.5N	97.6E
Shumkar, Pik see: Pik Shumkar				
Shunhe	PPL	C	28.7N	102.5E
Shurab	PPL	U	40.1N	70.6E
Shuroabad	PPL	U	37.8N	70.1E
Shürong	PPL	C	31.8N	89.7E
*Shyok	RIV	I	34.6N	77.4E
*Siachen Glacier	GLCR	P	35.4N	77.0E
*Siálkot	PPL	P	32.5N	74.5E
Sibda see: Sinda				
Sibsāgar	PPL	I	26.9N	94.7E
Sidzhak	PPL	U	41.7N	70.1E
Si'ergou	PPL	C	34.4N	104.1E
Si'ertan	PPL	C	37.2N	103.8E
Sihe	PPL	C	28.2N	100.8E
Sikai	PPL	C	27.9N	102.7E
*Sikar	PPL	C	27.6N	75.2E
*Silgarhi-Doti	PPL	N	29.3N	81.0E
Silghāt	PPL	I	26.6N	92.9E
*Siling Co	LK	C	31.7N	89.0E
Sima	PPL	C	31.9N	91.0E
Simankou	PPL	C	39.1N	99.8E
Simawat	PPL	C	37.7N	80.3E
Simen	PPL	C	34.6N	105.0E
Simeng	PPL	C	29.9N	103.7E
*Simikot	PPL	N	30.0N	81.8E

name	designation	area	lat.	long.
*Simla	PPL	I	31.1N	77.2E
Sinda	PPL	C	32.0N	97.5E
*Sindhūli Garhi	PPL	N	27.3N	86.0E
Singgimtay	PPL	C	42.9N	89.7E
Singim see: Singgimtay				
Singkaling Hkāmti see: Hkāmti				
Sipingchang	PPL	C	29.7N	102.6E
*Sirha	PPL	N	26.7N	86.2E
Sirsa	PPL	I	29.5N	75.1E
*Sitāmarhi	PPL	I	26.6N	85.5E
*Sitapur	PPL	I	27.6N	80.7E
Sitian	PPL	C	42.0N	94.5E
Siwālik Hills see: Churia Range				
*Siwān	PPL	I	26.2N	84.4E
Sixin	PPL	C	37.1N	99.7E
Siying	PPL	C	25.2N	103.1E
*Siyu He	RIV	C	27.6N	103.7E
Sizu	PPL	C	28.8N	102.7E
Skalistyy, Pik see: Pik Skalistyy				
*Skārdu	PPL	P	35.3N	75.6E
Skobeleva, Pik see: Pik Skobeleva				
Snezhnaya, Gora see: Gora Snezhnaya				
Soche see: Shache				
Sogat	PPL	C	36.3N	78.0E
Sogat	PPL	C	38.8N	76.2E
Sogat	PPL	C	41.4N	87.2E
Sogcanggoniba	PPL	C	33.4N	102.4E
Sogdoi	PPL	C	31.6N	80.3E
Sogety, Gory see: Gory Sogety				
*Sogma	PPL	C	34.5N	80.4E
Sogmai	PPL	C	32.2N	79.9E
Sognag see: Sonag				
Sogo Nur	PPL	C	42.2N	101.0E
*Sogo Nur	LK	C	42.3N	101.2E
Sog Qu	RIV	C	31.4N	93.9E

72

name	designation	area	lat.	long.
Taklakot see: Burang				
*Taklimakan Shamo	DES	C	38.1N	82.0E
Taktalyk, Khrebet see: Khrebet Taktalyk				
Talagang	PPL	P	32.9N	72.4E
*Talas	PPL	U	42.5N	72.2E
*Talas	RIV	U	42.5N	72.2E
Talasskiy Alatau, Khrebet see: Khrebet Talasskiy Alatau				
Taldyk	RIV	U	40.3N	73.3E
Taldysu see: Taldy-Suu				
Taldy-Suu	PPL	U	42.8N	78.5E
Talgar	PPL	U	43.3N	77.3E
Talgar	RIV	U	43.3N	77.3E
Talgar, Gora see: Gora Talgar				
Talgar, Pik see: Gora Talgar				
Talu	PPL	C	30.3N	94.6E
Tamgäs	PPL	N	28.1N	83.3E
Tamsag Bulag	PPL	C	40.2N	103.0E
Tamsag Muchang	PPL	C	40.2N	103.0E
*Tamur	RIV	N	27.1N	87.7E
*Tanaing	PPL	B	26.3N	96.8E
Tanakpur	PPL	I	29.1N	80.1E
Tända	PPL	I	26.5N	82.7E
Tangchang	PPL	C	30.9N	103.8E
Tangdan	PPL	C	26.1N	103.0E
Tangdê	PPL	C	30.8N	93.3E
Tange	PPL	C	34.6N	104.8E
Tanggar	PPL	C	29.8N	91.5E
Tanggar	PPL	C	30.5N	96.6E
Tanggarma	PPL	C	36.2N	99.8E
Tanggarma Nongchang	PPL	C	36.0N	100.0E
Tanggarmo see: Tanggarma Nongchang				
Tanggo	PPL	C	29.1N	101.4E
Tanggo	PPL	C	30.2N	91.5E
Tanggo	PPL	U	31.5N	93.0E
Tanggor	PPL	C	33.4N	102.4E

name	designation	area	lat.	long.
*Tanggula Shan	MTS	C	32.8N	91.0E
*Tanggula Shan	PK	C	33.2N	91.2E
Tanggulashan see: Tuotuoheyan				
*Tanggula Shankou	PASS	C	32.8N	91.9E
Tanggulashanqu see: Tuotuoheyan				
*Tanggyai	PPL	C	30.6N	94.2E
Tangjiahai	PPL	C	38.2N	103.2E
Tanglag	PPL	C	33.9N	99.4E
Tanglhai	PPL	C	31.0N	88.7E
Tanglha Range see: Tanggula Shan				
*Tangluqangma	PPL	C	32.0N	90.2E
Tangmai	PPL	C	30.1N	95.0E
Tangnag	PPL	C	35.5N	100.1E
Tangpu	PPL	C	31.5N	98.3E
Tangqungmai	PPL	C	31.6N	86.7E
*Tangra Yumco	LK	C	31.2N	86.7E
Tangtang	PPL	C	26.4N	102.7E
Tangwangchuan	PPL	C	35.8N	103.5E
Tangxung	PPL	C	33.9N	99.5E
Tangyang	PPL	C	28.7N	100.8E
Tangzi	PPL	C	25.4N	103.2E
*Taniantaweng Shan	MTS	C	29.0N	98.4E
Tanjiajing	PPL	I	37.6N	103.5E
Tankse	PPL	I	34.1N	78.2E
*Tansing	PPL	N	27.9N	83.6E
Tanymas	RIV	U	38.6N	72.8E
Taoju	PPL	C	25.3N	101.6E
Taoping	PPL	C	34.1N	104.9E
*Taplejung	PPL	N	27.4N	87.7E
Tar	RIV	U	40.6N	73.6E
Taragay	RIV	U	41.6N	77.9E
*Tarbela Reservoir	RSV	P	34.3N	72.8E
Targalak	PPL	C	39.4N	74.8E
Targap	PPL	U	43.3N	75.8E
Targyai	PPL	C	28.3N	88.5E

name	designation	area	lat.	long.
Targyailing	PPL	C	29.4N	85.0E
Tarim	PPL	C	41.0N	83.1E
Tarim Bachang	PPL	C	41.0N	86.4E
Tarim Erchang	PPL	C	40.6N	87.9E
Tarim Liuchang	PPL	C	40.7N	87.2E
*Tarim Pendi	BSN		39.1N	81.5E
Tarim Qichang	PPL	C	40.9N	86.5E
Tariskay Shan	PK		42.6N	82.6E
Tarlak	PPL	C	41.9N	84.2E
*Tarlha Ri	PK		28.3N	91.1E
Tarmar	PPL	C	29.4N	88.5E
*Taro Co	LK		31.1N	84.1E
*Tarrong	⊙PPL	C	29.4N	90.2E
Tarsumdo	PPL	C	33.7N	99.4E
Tart	PPL	C	37.0N	92.9E
Tashbulak	PPL	U	40.9N	71.6E
*Tashi	PPL	C	40.2N	95.9E
*Tashigang	PPL	Z	27.3N	91.6E
Tashigong see: Zhaxigang				
Tashi He	RIV	C	40.0N	95.9E
*Tashkent	PPL	U	41.3N	69.3E
*Tash-Kumyr	PPL	U	41.4N	72.2E
Tashlak	PPL	U	40.5N	71.8E
*Tatalin Gol	RIV	C	37.7N	96.4E
Tatirkotan	PPL	C	40.8N	82.2E
Tatlikbulak	PPL	C	38.8N	87.8E
Tatrang	PPL	C	38.5N	85.7E
*Tatu	PPL	N	35.6N	98.1E
Taulihawa	PPL	C	27.6N	83.1E
Tavil'dara	PPL	U	38.7N	70.5E
Tawa	PPL	C	32.5N	101.5E
Tawan	PPL	C	37.1N	101.2E
Tawang	PPL	I	27.6N	91.9E
Taxdian	PPL	C	41.8N	86.2E
*Taxkorgan	PPL	C	37.7N	75.2E

name	designation	area	lat.	long.
Tazang	PPL	C	33.3N	103.8E
Tazgun	PPL	C	37.3N	78.5E
Taziba	PPL	C	30.1N	100.5E
Tebh	PPL	C	36.3N	97.6E
Tébo	PPL	C	31.3N	90.3E
Tehri	PPL	I	30.4N	78.5E
*Teju	PPL	I	27.9N	96.2E
Tekes	RIV	U	42.8N	79.6E
Tekes	PPL	U	42.8N	80.1E
*Tekes	PPL	C	43.2N	81.8E
*Tekes He	RIV	C	43.0N	81.4E
Tekiliktag	PK	C	36.5N	80.4E
Telashi Hu	LK	C	34.8N	92.2E
Témarxung	PPL	C	32.2N	88.7E
Teme	PPL	C	37.3N	101.3E
*Temirlanovka	PPL	U	42.6N	69.3E
Temirlik	RIV	U	43.3N	79.3E
Temo	PPL	C	29.4N	94.5E
Tempung	PPL	C	37.1N	99.3E
Temtog	PPL	C	29.8N	97.6E
Tengger Els	PPL	C	37.7N	104.8E
*Tengger Feng	PK	C	43.1N	86.8E
Tengger Shamo	DES	C	38.2N	104.0E
Tengkye Dzong see: Dinggyê				
Teploklyuchenka	PPL	U	42.5N	78.5E
Terek	RIV	U	40.5N	75.9E
Terek	RIV	U	41.0N	74.0E
*Terek-Say	PPL	U	41.5N	71.2E
Terektau, Khrebet see: Khrebet Terektau				
*Tergun Daba Shan	MTS	C	38.3N	96.3E
Terhathum	PPL	N	27.1N	87.5E
Téring	PPL	C	30.9N	88.1E
Ters	RIV	U	41.6N	70.5E
Terskey-Alatau, Khrebet see: Khrebet Terskey-Ala-Too				
Terskey-Ala-Too, Khrebet see: Khrebet Terskey-Ala-Too				

76

name	designation	area	lat.	long.
Tunggar	PPL	C	29.0N	93.2E
Tung La	PASS		30.6N	95.2E
Tungru	PPL	C	34.0N	80.7E
*Tunuk	PPL	U	42.2N	74.0E
*Tuoheping Co	LK	C	34.1N	83.1E
Tuo Jiang	RIV	C	30.7N	104.5E
Tuojue	PPL	C	27.5N	102.8E
Tuole	PPL	C	38.8N	98.4E
Tuomugou	PPL	C	27.6N	102.4E
Tuotuo He	RIV	U	34.2N	91.5E
Tuotuohe see: Tuotuoheyan				
*Tuotuoheyan	PPL	C	34.2N	92.4E
Tuowu	PPL	C	28.8N	102.2E
Tuoyun	PPL	C	39.9N	75.5E
Tura	PPL	C	37.7N	86.4E
Turakurgan	PPL	U	41.0N	71.5E
Turbat	PPL	U	41.8N	69.6E
Turfan see: Turpan				
Turgen	RIV	U	43.1N	77.6E
Turgen'	PPL	U	43.4N	77.6E
Turkestanskiy Khrebet	MTS	U	39.5N	69.0E
*Turpan	PPL	C	42.9N	89.2E
*Turpan Pendi	DEPR	C	42.8N	89.3E
Turpan Zhan	PPL	C	43.1N	88.8E
Turgart, Pereval see: Turugart Shankou				
Turugart Shankou	PASS	X	40.6N	75.4E
Turumtaykul', Ozero see: Ozero Turumtaykul'				
Tuura-Suu	RIV	U	42.2N	76.3E
Tüwat	PPL	C	37.4N	79.7E
Tuwopu	PPL	C	38.8N	102.1E
Tuyabuguz	PPL	U	41.0N	69.3E
Tuyao	PPL	C	42.0N	92.2E
Tuyuk	PPL	U	43.1N	79.4E
Tuzhuchang	PPL	C	29.6N	103.8E
Tuzkol', Ozero see: Ozero Tuzkol'				

name	designation	area	lat.	long.
Tuzluk	PPL	C	36.5N	79.8E
Tyan'-Shan' see: Tien Shan				
Tyugël'-Say	PPL	U	42.0N	74.8E
Tyulek	RIV	U	41.9N	75.6E
Tyul'kubas	PPL	U	42.5N	70.3E
*Tyup	PPL	U	42.8N	78.3E
Tyup	RIV	U	42.8N	78.5E
Tyuyabuguz see: Tuyabuguz				
Tyuzashu, Pereval see: Pereval Tyuz-Ashuu				
Tyuz-Ashuu, Pereval see: Pereval Tyuz-Ashuu				
Uchchat, Khrebet see: Khrebet Uchchat				
Uch-Kël'	RIV	U	41.9N	78.9E
Uch-Korgon	PPL	U	40.3N	72.1E
Uch-Koshoy	RIV	U	42.5N	73.0E
Uchkuprik	PPL	U	40.5N	71.1E
Uchkurgan	PPL	U	41.1N	72.1E
Uch"yar	PPL	U	39.9N	71.1E
*Udhampur	PPL	I	32.9N	75.1E
Ugam	RIV	U	41.9N	70.1E
Ugamskiy Khrebet	MTS	U	42.0N	70.3E
Ugut see: Ugyut				
*Ugyut	PPL	U	41.4N	74.9E
Ulan	RIV	U	41.3N	76.9E
*Ulan Buh Shamo	DES	C	40.6N	106.8E
Ulanbuy Shuiku	RSV	C	43.7N	87.6E
Ulanlinggi	PPL	C	42.8N	87.3E
Ulan Mod	PPL	C	39.9N	104.9E
Ulan Suhai	PPL	C	41.2N	104.1E
Ulan Tohoi	PPL	C	40.8N	101.5E
*Ulan Ul Hu	LK	C	34.8N	90.5E
Ulan Ul Shan	MTS	C	34.8N	91.0E
Ulan Xian	CCN	C	36.9N	98.4E
Ulan Yaodong	PPL	C	38.9N	96.2E
Ulastai	PPL	C	42.9N	86.7E
Ulugbek	PPL	U	41.4N	69.4E

78

name	designation	area	lat.	long.
Uluqqat	PPL	C	29.9N	74.3E
Uluqqat see: Wuqia				
*Una	PPL	I	31.5N	76.3E
Unuli Horog	PPL	C	35.1N	91.9E
Uqmurawan	PPL	C	39.6N	75.9E
Uqturpan see: Wushi			41.2N	79.2E
Urai Lägnä see: Biling La				
Urdoi	PPL	C	31.9N	87.3E
Uri	PPL	I	34.1N	74.1E
Urmai	PPL	C	31.7N	87.6E
*Urru Co	LK	C	31.7N	87.9E
Urtaaul	PPL	U	41.2N	69.1E
*Urt Moron	PPL	C	36.9N	93.1E
Urt Moron He	RIV	C	37.1N	93.8E
*Ürümqi	PPL	C	43.8N	87.6E
Ushkän	PPL	A	37.0N	71.0E
Uspenovka	PPL	U	43.2N	74.4E
*Uttarkashi	PPL	I	30.8N	78.3E
Uxu	PPL	C	29.3N	85.4E
*Uxxaktal	PPL	C	42.2N	87.2E
Uxxarbax	PPL	C	37.4N	77.4E
Uychi	PPL	U	41.0N	71.8E
*Uzatag	PK	C	36.3N	83.0E
Uz-Bel', Pereval see: Uzbel Shankou				
*Uzbel Shankou	PASS	X	38.6N	73.9E
Uzengegush, Gora see: Gora Uzëngyu-Kuush				
Uzëngyu-Kuush	RIV	U	41.2N	77.6E
Uzëngyu-Kuush, Gora see: Gora Uzëngyu-Kuush				
*Uzgen	PPL	U	40.8N	73.3E
Uzgenskiy Khrebet	MTS	U	40.7N	74.0E
Uzunagach	PPL	U	43.2N	76.3E
Uzunakhmat see: Uzun-Akmat				
Uzun-Akmat	RIV	U	42.1N	72.3E
*Uzun Bulak	SPNG	C	42.0N	88.9E
Uzunbulak	PPL	U	43.2N	79.0E
Uzunsay	PPL	C	37.0N	78.8E
Uzynbulak see: Uzunbulak				
Väkhän	RGN	A	37.0N	73.0E
Väkhän, Ab-e see: Ab-e Väkhän				
Vakhanskiy Khrebet	MTS	X	37.0N	73.7E
Väkhjir, Daryä-ye see: Daryä-ye Väkhjir				
Väkhjir, Kowtal-e see: Wakhjir Pass				
*Vakhsh	RIV	U	38.7N	69.7E
Vakhshskiy Khrebet	MTS	U	38.5N	69.5E
Vanch	PPL	U	38.4N	71.4E
Vanch	RIV	U	38.5N	71.5E
Vanchskiy Khrebet	MTS	U	38.4N	71.8E
*Vannovka	PPL	U	42.5N	70.3E
Varang see: Vrang				
Vardüj, Daryä-ye see: Daryä-ye Vardüj				
Viskharv	PPL	U	38.6N	71.1E
Vod Åb	PPL	A	38.3N	70.9E
Vorukh	PPL	U	39.8N	70.6E
Vostochnyy	PPL	U	39.9N	69.6E
Vostochnyy Karakol	RIV	U	42.4N	75.1E
Vostochnyy Sëok	RIV	U	42.3N	75.1E
Vostochnyy Suyek see: Vostochnyy Sëok				
Vrang	PPL	U	37.0N	72.4E
Vtoraya Pyatiletka	PPL	U	43.2N	76.9E
Vudor, Pik see: Pik Vudor				
Vysokoye	PPL	U	42.5N	70.5E
Wafang	PPL	C	25.3N	99.0E
Wafang	PPL	C	26.1N	100.7E
Wagang	PPL	C	27.9N	103.3E
Wahei	PPL	C	28.7N	103.1E
Wainapu	PPL	C	33.3N	105.1E
Wakhan Corridor see: Väkhän				
*Wakhjir Pass	PASS	X	37.1N	74.5E
Wa-k'o-chi-erh Shan-k'ou see: Wakhjir Pass				
Walêg	PPL	C	28.1N	101.5E

name	designation	area	lat.	long.
Wolongshi	PPL	C	30.0N	101.2E
Woring	PPL	A	37.3N	70.5E
Wozhang Shan	PK	C	25.8N	102.3E
Wubairha	PPL	C	41.9N	106.4E
Wubao	PPL	C	29.2N	104.4E
Wuchuan	PPL	C	36.7N	104.0E
Wudaoliang	PPL	C	35.1N	93.0E
Wuding Xian	CCN	C	25.5N	102.3E
Wudu	PPL	C	31.8N	104.7E
*Wudu	PPL	C	33.4N	104.9E
Wudun	PPL	C	40.1N	94.8E
Wufengxi	PPL	C	30.6N	104.4E
Wufosi	PPL	C	37.2N	104.3E
*Wüjang	PPL	C	33.6N	79.8E
Wuji	PPL	C	28.1N	103.5E
Wujiachuan	PPL	C	36.6N	104.5E
Wujiangpu	PPL	C	39.0N	100.4E
Wujing	PPL	C	27.6N	99.5E
Wulang	PPL	C	29.3N	95.2E
*Wular Lake	LK	I	34.3N	74.5E
*Wuli	PPL	C	34.4N	92.7E
Wulian Feng	PK	C	27.8N	103.6E
*Wulian Feng	MTS	C	28.0N	103.8E
Wulian He	RIV	C	28.8N	100.7E
Wulong	PPL	C	26.0N	103.1E
Wumatang	PPL	C	30.5N	91.4E
Wumeng	PPL	C	26.0N	102.7E
*Wungda	PPL	C	31.8N	100.7E
Wupo	PPL	C	28.1N	103.0E
*Wuqia	PPL	C	39.7N	75.1E
*Wushan	PPL	C	34.7N	104.8E
Wushaoling	PASS	C	37.2N	102.8E
Wushao Ling	MTS	C	37.2N	103.1E
Wushi	PPL	C	36.7N	102.1E
*Wushi	PPL	C	41.2N	79.2E

name	designation	area	lat.	long.
Wusihe	PPL	C	29.2N	102.8E
Wusituobie	PPL	C	38.5N	95.8E
Wutonggou	PPL	C	40.7N	98.6E
Wutonggou	PPL	C	41.9N	89.8E
Wutongqiao	PPL	C	29.4N	103.8E
Wutongwozi Quan	SPNG	C	42.4N	95.0E
*Wutuo Jing	WELL	C	39.4N	103.8E
Wuwei Xian	CCN	C	37.9N	102.6E
Wuxu	PPL	C	29.1N	101.4E
*Wuyang	⊙PPL	C	30.4N	103.8E
Wuyi	PPL	C	27.5N	102.9E
*Wuzhong	PPL	C	37.9N	106.2E
Xab Qu	RIV	C	29.2N	88.3E
Xabrang	PPL	C	35.5N	101.9E
Xabyai	PPL	C	30.8N	96.2E
Xabyaisamba	PPL	C	31.0N	96.2E
*Xaga	PPL	C	28.7N	85.7E
Xagar	PPL	C	30.5N	86.3E
Xagdomba	PPL	C	34.1N	102.6E
Xagjang	PPL	C	29.0N	92.2E
Xagnag	PPL	C	32.6N	81.5E
Xago	PPL	C	30.9N	88.3E
Xaguka	PPL	C	31.8N	92.8E
Xagqü Qu	RIV	C	31.1N	92.8E
Xahbulak	PPL	C	42.2N	83.2E
Xaidulla	PPL	C	36.4N	77.9E
Xainza	PPL	C	30.9N	88.6E
Xainza Xian	CCN	C	30.9N	88.6E
Xaitongmoin Xian	CCN	C	29.4N	88.2E
*Xakur	PPL	C	39.8N	79.4E
Xalatang	PPL	C	31.2N	100.7E
Xalazakung	PPL	C	31.0N	81.0E
Xamal	PPL	C	39.6N	78.4E
*Xamardêqên	PK	C	32.8N	93.2E
Xang	PPL	C	35.9N	97.9E

name	designation	area	lat.	long.
*Xiaohaizi Shuiku	RSV	C	39.7N	78.7E
Xiaohe	PPL	C	26.8N	103.0E
Xiaohe	PPL	C	32.5N	104.1E
Xiaohongshan	PPL	C	40.8N	99.2E
*Xiao Jiang	RIV	C	26.3N	103.1E
Xiaojin Chuan	RIV	C	31.0N	102.2E
*Xiaojin He	RIV	C	27.8N	101.4E
Xiaojin Xian	CCN	C	30.9N	102.3E
*Xiaonanchuan	PPL	C	35.7N	94.3E
*Xiaopuzi	PPL	C	27.5N	103.7E
Xiao Qaidam	PPL	C	37.5N	95.5E
Xiao Qaidam Hu	LK	C	37.4N	95.6E
Xiaoquan	PPL	C	31.2N	104.2E
Xiaoquandong	RIV	C	41.2N	95.7E
Xiaoshalong	PPL	C	38.8N	98.9E
Xiaoshao	PPL	C	25.1N	102.9E
Xiao Surmang	PPL	C	32.4N	97.2E
Xiaowan	PPL	C	40.5N	96.1E
*Xiaoxiang Ling	MTS	C	28.5N	102.5E
Xiaoxiang Shuiku	RSV	C	25.4N	103.7E
Xiaoyuan	PPL	C	29.9N	104.8E
Xiaozhongdian	PPL	C	27.5N	99.7E
Xiaqiaotou	PPL	C	27.1N	100.0E
Xiari'aba	PPL	C	32.6N	94.1E
*Xiari'aba Shan	PK	C	32.6N	94.1E
Xiaruo	PPL	C	27.7N	99.2E
Xiasifen	PPL	C	38.6N	102.2E
Xiaxihao	PPL	C	40.3N	97.1E
Xia Zanggor	PPL	C	33.9N	100.8E
Xia Zayü	PPL	C	28.4N	97.0E
Xiazhai	PPL	C	31.6N	102.0E
Xiazhuang	PPL	C	25.4N	100.8E
Xibdê	PPL	C	28.7N	99.8E
Xichahe	PPL	C	38.9N	99.3E
*Xichang	PPL	C	27.8N	102.2E

name	designation	area	lat.	long.
Xichuan He	RIV	C	36.7N	101.3E
*Xida Shan	PK	C	41.4N	88.2E
Xidatan	PPL	C	38.8N	106.3E
Xide Xian	CCN	C	28.3N	102.4E
Xidong	PPL	C	39.6N	98.4E
Xiehe	PPL	C	37.7N	102.7E
*Xigazê	PPL	C	29.2N	88.8E
Xi Golog	PPL	C	30.0N	100.2E
Xigongyi	PPL	C	35.6N	104.8E
Xihan Shui	RIV	C	33.8N	105.2E
Xihe	PPL	C	29.0N	103.0E
*Xihe	PPL	C	34.0N	105.2E
Xi He	RIV	C	41.4N	100.4E
Xihe Shuiku	RSV	C	25.6N	103.7E
Xihu	PPL	C	40.5N	95.0E
*Xiji	PPL	C	36.0N	105.7E
*Xijian Quan	SPNG	C	40.8N	96.6E
Xijir	PPL	C	34.8N	91.9E
Xijir	PPL	C	43.9N	90.0E
*Xijir Ulan Hu	LK	C	35.2N	90.4E
Xijishui	PPL	C	37.0N	104.0E
Xiliangzi	PPL	C	38.9N	93.8E
⊙Xiligou	⊙PPL	C	36.9N	98.4E
Xiligou Hu	LK	C	36.8N	98.4E
Xilizhai	PPL	C	31.3N	102.1E
Xilong	PPL	C	28.2N	102.1E
Xiluo	PPL	C	27.6N	102.6E
Ximiao	PPL	C	41.0N	100.3E
Xinba	PPL	C	39.1N	99.3E
Xinbao	PPL	C	34.6N	103.6E
Xinbao	PPL	C	34.8N	103.8E
Xinchang	PPL	C	30.5N	103.3E
Xincheng	PPL	C	34.6N	103.9E
Xincheng	PPL	C	36.1N	103.4E
Xincheng	PPL	C	38.2N	101.5E

name	designation	area	lat.	long.
Xincun	PPL	C	26.6N	102.9E
Xincun see: Dongchuan				
Xindian	PPL	C	26.1N	103.2E
Xindian	PPL	C	25.2N	101.5E
Xindian	PPL	C	30.1N	103.2E
Xindian	PPL	C	35.6N	103.7E
Xindianzi	PPL	C	27.6N	103.5E
Xindianzi	PPL	C	30.9N	103.0E
*Xindu	PPL	C	30.8N	104.1E
*Xinduqiao	PPL	C	30.1N	101.6E
Xinfan	PPL	C	30.8N	104.0E
Xingba	PPL	C	30.2N	81.2E
Xingdi	PPL	C	41.2N	87.9E
Xinghai Xian	CCN	C	35.5N	100.0E
Xinghu	PPL	C	26.4N	100.7E
Xinglongzhen	PPL	C	34.4N	105.7E
Xingqêngoin	PPL	C	31.9N	100.2E
*Xingrenbu	PPL	C	36.9N	105.2E
Xingrong	PPL	C	30.9N	95.7E
*Xingsagoinba	PPL	C	34.2N	101.5E
Xingta	PPL	C	31.4N	104.4E
Xingxingxia	PPL	C	41.8N	95.1E
Xingxiu Hai	LK	C	35.0N	96.7E
*Xinhe	PPL	C	41.6N	82.6E
Xin Hu	LK	C	34.4N	84.2E
Xinhua	PPL	C	29.2N	103.5E
Xinhua	PPL	C	39.2N	100.0E
Xinhuacun	PPL	C	36.9N	103.1E
Xining	PPL	C	28.5N	103.6E
*Xining	PPL	C	36.6N	101.8E
Xinjie	PPL	C	25.0N	99.1E
Xinjie	PPL	C	25.0N	99.5E
Xinjie	PPL	C	25.8N	101.2E
Xinjie	PPL	C	26.8N	102.7E
Xinjie	PPL	C	35.7N	101.3E
Xinjin Xian	CCN	C	30.4N	103.8E
Xinlian	PPL	C	28.5N	99.6E
*Xinlong Xian	CCN	C	30.9N	100.3E
Xinluhai	PPL	C	31.8N	99.0E
Xinmin	PPL	C	28.8N	103.7E
Xinmin	PPL	C	29.3N	102.2E
Xinmin	PPL	C	29.3N	102.5E
Xinmincun	PPL	C	27.2N	103.6E
Xinminpu	PPL	C	39.9N	97.8E
Xinpuzi	PPL	C	37.2N	103.7E
Xinqiao	PPL	C	29.6N	103.8E
Xinsanchaba	PPL	C	30.3N	104.2E
*Xinshiba	⊙PPL	C	28.9N	102.7E
Xinshizhen	PPL	C	28.6N	103.8E
Xinsi	PPL	C	34.6N	104.6E
Xinsi	PPL	C	36.0N	103.0E
Xintianpu	PPL	C	35.5N	103.8E
Xintun	PPL	C	26.6N	100.2E
Xintunchuan	PPL	C	36.3N	103.4E
Xinxing	PPL	C	31.1N	103.8E
Xinyangzhen	PPL	C	34.6N	105.5E
Xinying	PPL	C	35.7N	104.1E
Xinying	PPL	C	36.0N	105.6E
*Xinyuan	⊙PPL	C	37.3N	99.0E
*Xinyuan	PPL	C	43.4N	83.2E
Xipu	PPL	C	30.7N	103.9E
Xiqiao	PPL	C	24.9N	103.6E
*Xiqing Shan	MTS	C	34.6N	101.5E
Xiqu	PPL	C	38.9N	103.5E
Xireg see: Xiligou				
Xiri	PPL	C	32.0N	103.1E
Xirong	PPL	C	29.4N	103.7E
Xishan	PPL	C	25.0N	102.4E
Xi Taijnar Hu	LK	C	37.6N	93.4E
Xitie Shan	MTS	C	37.2N	95.8E
Xitieshan	PPL	C	37.3N	95.6E

name	designation	area	lat.	long.
Xiwu	PPL	C	33.1N	97.3E
*Xixabangma Feng	PK		28.3N	85.7E
Xixi He	RIV	C	27.7N	103.0E
Xiyanchi	PPL	C	43.4N	91.0E
Xiyao	PPL	C	25.0N	100.2E
Xiyêrkang	MTS	C	33.1N	90.5E
Xiying	PPL	C	37.9N	102.3E
Xiying He	RIV	C	37.9N	102.1E
Xize	PPL	C	26.2N	103.8E
Xizêkar see: Gongqên				
Xizhou	PPL	C	25.8N	100.1E
Xobando	PPL	C	30.8N	95.6E
Xognga	PPL	C	31.8N	96.2E
*Xoi	⊙PPL	C	29.3N	90.7E
*Xoilapu Kangri	PK	C	29.7N	88.2E
Xoima	PPL	C	34.9N	100.6E
Xoisar	PPL	C	28.6N	92.5E
Xoka	PPL	C	29.9N	93.8E
Xolo	PPL	C	28.2N	100.6E
Xolungtang	PPL	C	37.1N	102.1E
Xo Qu	RIV	C	28.5N	99.5E
Xordang	PPL	C	28.4N	98.8E
*Xorkol	PPL	C	38.7N	91.0E
Xorkol	PPL	C	39.2N	87.5E
Xortang	PPL	C	37.8N	84.1E
Xowa	PPL	C	33.5N	103.8E
Xuanhepu	PPL	C	37.4N	105.5E
Xuankou	PPL	C	30.9N	103.5E
*Xuebang Shan	PK	C	26.5N	99.5E
*Xuebao Ding	PK	C	32.6N	103.8E
Xuecheng	PPL	C	31.5N	103.3E
Xueheli	PPL	C	38.7N	98.0E
Xuehuan Hu	LK	C	35.0N	88.0E
Xuejing Hu	LK	C	35.9N	87.3E
Xuepan Shan	MTS	C	26.2N	99.2E

name	designation	area	lat.	long.
*Xue Shan	MTS	C	27.4N	99.7E
Xueshan	PPL	C	31.4N	102.5E
Xueshan	PPL	C	34.8N	99.7E
*Xugin Gol He	RIV	C	35.7N	95.6E
Xugui	PPL	C	35.7N	95.9E
Xugui	PPL	C	40.2N	103.8E
Xujie	PPL	C	25.5N	101.6E
Xümatang	PPL	C	33.8N	97.3E
Xümo	PPL	C	30.2N	95.3E
*Xundian	PPL	C	25.5N	103.2E
Xungba	PPL	C	30.8N	81.1E
*Xungba	PPL	C	32.0N	81.9E
*Xungmai	PPL	C	31.3N	89.0E
Xungqên	PPL	C	30.9N	90.1E
Xung Qu	RIV	C	28.4N	92.2E
Xungru	PPL	C	29.2N	84.7E
Xunhua Salarzu Zizhixian	CCN	C	35.8N	102.4E
Xunsichang	PPL	C	28.0N	104.5E
Xur	PPL	C	35.7N	95.6E
Xurgan	PPL	C	35.7N	95.6E
*Xuru Co	LK	C	30.3N	86.4E
Xusanwan	PPL	C	39.3N	99.3E
Xushuizhen	PPL	C	31.5N	104.3E
Xutangpu	PPL	C	32.2N	104.4E
*Ya'an	PPL	C	30.0N	102.9E
Yabrai	PPL	C	39.7N	103.0E
*Yabrai Shan	MTS	C	39.8N	103.0E
Yabrai Yanchang	PPL	C	39.3N	102.9E
Yabut	PPL	C	42.3N	101.0E
Yacheng	PPL	C	34.3N	105.1E
Yadong Xian	CCN	C	27.4N	88.9E
*Yagan	PPL	C	42.0N	102.6E
*Yaggain Canco	LK	C	32.9N	89.8E
Yagmo	PPL	C	29.4N	87.6E
Yagmo	PPL	C	29.7N	87.8E

name	designation	area	lat.	long.
Zizhong	PPL	C	29.8N	104.8E
Zoco	PPL	C	32.3N	80.4E
Zogainrawar see: Huashixia				
Zogang Xian	CCN	C	29.6N	97.8E
Zogqên	PPL	C	32.1N	98.8E
Zoidê Lhai	PPL	C	32.5N	88.1E
*Zoigê'nyinma	⊙PPL	C	34.0N	102.1E
Zoigê Xian	CCN	C	33.5N	102.9E
Zomdo	PPL	C	31.1N	99.0E
*Zonag Co	LK	C	35.5N	91.9E
*Zong	PPL	U	37.1N	72.6E
Zongbur	PPL	C	32.2N	101.5E
*Zongga	⊙PPL	C	28.9N	85.2E
Zonggag	PPL	C	31.9N	102.1E
*Zongjia	PPL	C	36.2N	97.0E
Zongjiafangzi	PPL	C	36.3N	97.0E
Zonglung	PPL	C	32.3N	99.5E
Zongxoi	PPL	C	29.8N	91.2E
Zongza	PPL	C	29.3N	99.3E
Zongzhai	PPL	C	36.5N	101.6E
Zongzhai	PPL	C	39.6N	98.6E
Zorkul', Ozero see: Ozero Zorkul'				
Zoulang Nanshan	MTS	C	38.5N	99.5E
*Zuli He	RIV	C	36.4N	104.8E
Zulumart, Khrebet see: Khrebet Zulumart				
Zuosuo	PPL	C	27.7N	100.8E

LISTING OF CERTAIN CHINESE COUNTY AND COUNTY CENTRE NAMES

Aksay Kazakzu Zizhixian: centre is Bolozhuanjing
Alxa Youqi: centre is Ehen Hudag
Alxa Zuoqi: centre is Bayan Hot
Amdo Xian: centre is Pagnag
Anyuan: centre of Tianzhu Zangzu Zizhixian

Babao: centre of Qilian Xian
Baima: centre of Baxoi Xian
Baima Xian: centre is Séraitang
Bainang Xian: centre is Norbukyungzê
Baingoin Xian: centre is Pubu

Baiyü Xian: centre is in Jianshe commune
Banbar Xian: centre is Damartang
Baohe: centre of Weixi Xian
Baqên Xian: centre is Dartang
Batang Xian: centre is in Qianjin commune

Baxoi Xian: centre is Baima
Bayan: centre of Hualong Huizu Zizhixian
Bayan Hot: centre of Alxa Zuoqi
Bijiang Xian: centre is Zhiziluo
Binchuan Xian: centre is Niujing

Bohu Xian: centre is Boron Sum
Bolo: centre of Gonjo Xian
Bolozhuanjing: centre of Aksay Kazakzu Zizhixian
Bomi Xian: centre is Zhamo
Boron Sum: centre of Bohu Xian

Bowa: centre of Muli Zangzu Zizhixian
Butuo Xian: centre is in Temuli commune
Chatang: centre of Zhanang Xian
Chindu Xian: centre is Chuqung
Chugqênsumdo: centre of Jigzhi Xian

Chuimatan: centre of Jishishan Bonanzu Dongxiangzu Salarzu Zizhixian
Chumba: centre of Gyaca Xian
Chuqung: centre of Chindu Xian
Comai Xian: centre is Damxoi
Congdü: centre of Nyalam Xian

Coqên Xian: centre is Maindong
Dagcagoin: centre of Zoigê Xian
Dagma: centre of Lhari Xian
Dagzê Xian: centre is Dêqên
Dalain Hob: centre of Ejin Qi

Damquka: centre of Damxung Xian
Damxoi: centre of Comai Xian
Damxung Xian: centre is Damquka
Dangchengwan: centre of Subei Mongolzu Zizhixian
Daocheng Xian: centre is in Rubu commune

Darlag Xian: centre is Gyümai
Dartang: centre of Baqên Xian
Datong Xian: centre is Qiaotou
Dawu: centre of Maqên Xian
Dawukou: centre of Shizuishan Shi

Daxing: centre of Ninglang Yizu Zizhixian
Dayan: centre of Lijiang Naxizu Zizhixian
Dayao Xian: centre is Jinbi
Dêgê Xian: centre is in Goinqên commune
Dêngkagoin: centre of Têwo Xian

Dêngqên Xian: centre is Gyamotang
Dêqên: centre of Dagzê Xian
Dêqên Xian: centre is in Gaofeng commune
Dêrong Xian: centre is in Songmai commune
Dinggyê Xian: centre is Gyangkar

Doilungdêqên Xian: centre is Namka
Domartang: centre of Banbar Xian
Dongchuan: centre of Yao'an Xian
Dongxiangzu Zizhixian: centre is Suonanba
Dulan Xian: centre is Qagan Us

Ehen Hudag: centre of Alxa Youqi
Ejin Qi: centre is Dalain Hob
Fangting: centre of Shifang Xian
Fengyi: centre of Maowen Qiangzu Zizhixian
Fugong Xian: centre is in Shangpa commune
Fulin: centre of Hanyuan Xian
Gabasumdo: centre of Tongde Xian
Gadê Xian: centre is Pagqên
Gamda: centre of Zamtang Xian
Ganga Xian: centre is Shaliuhe
Ganluo Xian: centre is Xinshiba
Ganxiangying: centre of Xide Xian
Ganzhou: centre of Zhangye Xian
Gaolan Xian: centre is Shidongsi
Garbo: centre of Lhozhag Xian
Gartog: centre of Markam Xian
Gêding: centre of Xaitongmoin Xian
Gê'gyai Xian: centre is Napug
Gêrzê Xian: centre is Lumaringbo
Golingka: centre of Gongbo'gyamda Xian
Gongbo'gyamda Xian: centre is Golingka
Gonggar Xian: centre is Ramai
Gonghe Xian: centre is Qabqa
Gongshan Drungzu Nuzu Zizhixian: centre is in Cikai commune
Gonjo Xian: centre is Bolo
Guanghe Xian: centre is Taizisi
Guankou: centre of Guan Xian
Guan Xian: centre is Guankou
Guide Xian: centre is Heyin
Guinan Xian: centre is Mangra
Gyaca Xian: centre is Chumba
Gya'gya: centre of Saga Xian
Gyaijêpozhanggê: centre of Zhidoi Xian
Gyamotang: centre of Dêngqên Xian
Gyangkar: centre of Dinggyê Xian

Gyêgu: centre of Yushu Xian
Gyigang: centre of Zayü Xian
Gyirong Xian: centre is Zongga
Gyitang: centre of Lhünzê Xian
Gyümai: centre of Darlag Xian
Haiyan Xian: centre is Sanjiaocheng
Hanyuan Xian: centre is Fulin
Haomen: centre of Menyuan Huizu Zizhixian
Heishui Xian: centre is Luhua
Henan Mongolzu Zizhixian: centre is Yêgainnyin
Heyin: centre of Guide Xian
Hongwan: centre of Sunan Yugurzu Zizhixian
Hongyuan Xian: centre is Hurama
Hualong Huizu Zizhixian: centre is Bayan
Huangzhong Xian: centre is Lushar
Huaping Xian: centre is Zhongxin
Huidong Xian: centre is in Qianjin commune
Hurama: centre of Hongyuan Xian
Huzhu Tuzu Zizhixian: centre is Weiyuan
Jainca Xian: centre is Magitang
Jianchuan Xian: centre is Jinhuazhen
Jigzhi Xian: centre is Chugqênsumdo
Jin'an: centre of Songpan Xian
Jinbi: centre of Dayao Xian
Jincheng: centre of Wuding Xian
Jingtai Xian: centre is Yitiaoshan
Jingxin: centre of Yongshan Xian
Jinhuazhen: centre of Jianchuan Xian
Jinkouhe Gongnongqu: centre is in Heping commune
Jintang Xian: centre is Zhaozhen
Jinyang Xian: centre is in Tiantai commune
Jishi: centre is of Xunhua Salarzu Zizhixian
Jishishan Bonanzu Dongxiangzu Salarzu Zizhixian: centre is Chuimatan
Jiucheng: centre of Lintan Xian
Jiuquan Xian: centre is Suzhou

Kunggar: centre of Maizhokunggar Xian
Labrang: centre of Xiahe Xian
Lanping Xian: centre is in Lajing commune
Ledu Xian: centre is Nianbai
Lhari Xian: centre is Dagma

Lhazê Xian: centre is Quxar
Lhorong Xian: centre is Zito
Lhozhag Xian: centre is Garbo
Lhünzê Xian: centre is Gyitang
Lhünzhub Xian: centre is Poindo

Liangzhou: centre of Wuwei Xian
Lijiang Naxizu Zizhixian: centre is Dayan
Lintan Xian: centre is Jiucheng
Linze Xian: centre is Shahepu
Litang Xian: centre is in Gaocheng commune

Li Xian: centre is Zagunao
Ludian Xian: centre is Wenping
Luding Xian: centre is Luqiao
Luhua: centre of Heishui Xian
Luhuo Xian: centre is in Laojie commune

Lumaringbo: centre of Gêrzê Xian
Luqiao: centre of Luding Xian
Luquan Xian: centre is Pingshan
Luqu Xian: centre is Qiaotou
Lushan Xian: centre is in Luyang commune

Lushar: centre of Huangzhong Xian
Lushui Xian: centre is in Luzhang commune
Machali: centre of Madoi Xian
Madoi Xian: centre is Machali
Magitang: centre of Jainca Xian

Maindong: centre of Coqên Xian
Mainling Xian: centre is Tungdor
Maizhokunggar Xian: centre is Kunggar
Malong Xian: centre is Tongquan
Mangra: centre of Guinan Xian

Maowen Qiangzu Zizhixian: centre is Fengyi
Maqên Xian: centre is Dawu
Maqu Xian: centre is Zoigê'nyinma
Markam Xian: centre is Gartog
Meigu Xian: centre is in Bapu commune

Meixing: centre of Xiaojin Xian
Menyuan Huizu Zizhixian: centre is Haomen
Minhe Xian: centre is Shangchuankou
Muli Zangzu Zizhixian: centre is Bowa
Namka: centre of Doilungdêqên Xian

Nangqên Xian: centre is Xangda
Napug: centre of Gê'gyai Xian
Nianbai: centre of Ledu Xian
Ninglang Yizu Zizhixian: centre is Daxing
Ningnan Xian: centre is in Dongfeng commune

Niujing: centre of Binchuan Xian
Norbukyungzê: centre of Bainang Xian
Nyainrong Xian: centre is Sêrkang
Nyalam Xian: centre is Congdü
Nyêmo Xian: centre is Tarrong

Nyingchi Xian: centre is Pula
Pagnag: centre of Amdo Xian
Pagqên: centre of Gadê Xian
Peng Xian: centre is Tianpeng
Ping'an Xian: centre is Ping'anyi

Ping'anyi: centre of Ping'an Xian
Pingshan: centre of Luquan Xian
Pitong: centre of Pi Xian
Pi Xian: centre is Pitong
Poindo: centre of Lhünzhub Xian

Pubu: centre of Baingoin Xian
Puge Xian: centre is in Pule commune
Pula: centre of Nyingchi Xian
Qabqa: centre of : centre of Gonghe Xian
Qagan Us: centre of Dulan Xian

Oajortêbu: centre of Xainza Xian
Qamqên: centre of Rinbung Xian
Qiaotou (34.6N 102.4E): centre of Luqu Xian
Qiaotou (36.9N 101.6E): centre of Datong Xian
Qilian Xian: centre is Babao

Qingtongxia Xian: centre is Xiaoba
Qumarlêb Xian: centre is Yecangtan
Quxar: centre of Lhazê Xian
Qüxü Xian: centre is Xoi
Racaka: centre of Riwoqê Xian

Ramai: centre of Gonggar Xian
Rinbung Xian: centre is Qamqên
Riwoqê Xian: centre is Racaka
Rongwo: centre of Tongren Xian
Saga Xian: centre is Gya'gya

Sa'gya Xian: centre is Sa'gyaxoi
Sa'gyaxoi: centre of Sa'gya Xian
Sanjiaocheng: centre of Haiyan Xian
Sêraitang: centre of Baima Xian
Sêrkang: centre of Nyainrong Xian

Sêrtar Xian: centre is in Seke commune
Sêrxü Xian: centre is in Qümong commune
Shahepu: centre of Linze Xian
Shaliuhe: centre of Ganga Xian
Shangchuankou: centre of Minhe Xian

Shangjie: centre of Yangbi Xian
Shidongsi: centre of Gaolan Xian
Shifang Xian: centre is Fangting
Shimen: centre of Yunlong Xian
Shizuishan Shi: centre is Dawukou

Sog Xian: centre is Zaindainxoi
Sonag: centre of Zêkog Xian
Songming Xian: centre is Songyangzhen
Songpan Xian: centre is Jin'an
Songyangzhen: centre of Songming Xian

Subei Mongolzu Zizhixian: centre is Dangchengwan
Suijiang Xian: centre is Zhongcheng
Sunan Yuguzu Zizhixian: centre is Hongwan
Suonanba: centre of Dongxiangzu Zizhixian
Suzhou: centre of Jiuquan Xian

Taizisi: centre of Guanghe Xian
Tarrong: centre of Nyêmo Xian
Têwo Xian: centre is Dêngkagoin
Tianjun Xian: centre is Xinyuan
Tianpeng: centre of Peng Xian

Tianzhu Zangzu Zizhixian: centre is Anyuan
Tingri Xian: centre is Xêgar
Toling: centre of Zanda Xian
Tongde Xian: centre is Gabasumdo
Tongquan: centre of Malong Xian

Tongren Xian: centre is Rongwo
Tungdor: centre of Mainling Xian
Ulan Xian: centre is Xiligou
Wangda: centre of Zogang Xian
Weixi Xian: centre is Baohe

Weiyuan: centre of Huzhu Tuzu Zizhixian
Weizhou: centre of Wenchuan Xian
Wenchuan Xian: centre is Weizhou
Wenping: centre of Ludian Xian
Wuding Xian: centre is Jincheng

Wuwei Xian: centre is Liangzhou
Wuyang: centre of Xinjin Xian
Xainza Xian: centre is Qajortêbu
Xaitongmoin Xian: centre is Gêding
Xangda: centre of Nangqên Xian

Xarsingma: centre of Yadong Xian
Xêgar: centre of Tingri Xian
Xiahe Xian: centre is Labrang
Xiangcheng Xian: centre is in Hexi commune
Xiaoba: centre of Qingtongxia Xian

Xiaochuan: centre of Yongjing Xian
Xiaojin Xian: centre is Meixing
Xide Xian: centre is Ganxiangying
Xiligou: centre of Ulan Xian
Xinghai Xian: centre is Ziketan

Xinjin Xian: centre is Wuyang
Xinlong Xian: centre is in Wuxi commune
Xinshiba: centre of Ganluo Xian
Xinyuan: centre of Tianjun Xian
Xoi: centre of Qüxü Xian

Xunhua Salarzu Zizhixian: centre is Jishi
Yadong Xian: centre is Xarsingma
Yajiang Xian: centre is in Hekou commune
Yangbi Xian: centre is Shangjie
Yanghe: centre of Yongning Xian

Yanyuan Xian: centre is in Yanjing commune
Yao'an Xian: centre is Dongchuan
Yecangtan: centre of Qumarlêb Xian
Yêgainnyin: centre of Henan Mongolzu Zizhixian
Yêndum: centre of Zhag'yab Xian

Yitiaoshan: centre of Jingtai Xian
Yongding: centre of Yongren Xian
Yongjing Xian: centre is Xiaochuan
Yongning Xian: centre is Yanghe
Yongren Xian: centre is Yongding

Yongshan Xian: centre is Jingxin
Yuanma: centre of Yuanmou Xian
Yuanmou Xian: centre is Yuanma
Yunlong Xian: centre is Shimen
Yushu Xian: centre is Gyêgu

Yuyuriben: centre of Zadoi Xian
Zadoi Xian: centre is Yuyuriben
Zagunao: centre of Li Xian
Zaindainxoi: centre of Sog Xian
Zamtang Xian: centre is Gamda

Zanda Xian: centre is Toling
Zayü Xian: centre is Gyigang
Zêkog Xian: centre is Sonag
Zhabdün: centre of Zhongba Xian
Zhag'yab Xian: centre is Yêndum

Zhamo: centre of Bomi Xian
Zhanang Xian: centre is Chatang
Zhangye Xian: centre is : centre is Ganzhou
Zhaozhen: centre of Jintang Xian
Zhidoi Xian: centre is Gyaijêpozhanggê

Zhiziluo: centre of Bijiang Xian
Zhongba Xian: centre is Zhabdün
Zhongcheng: centre of Suijiang Xian
Zhongdian Xian: centre is Zhongxin
Zhongxin (26.6N 101.2E): centre of Huaping Xian

Zhongxin (27.7N 99.7E): centre of Zhongdian Xian
Ziketan: centre of Xinghai Xian
Zito: centre of Lhorong Xian
Zoigê'nyinma: centre of Maqu Xian
Zoigê Xian: centre is Dagcagoin

Zogang Xian: centre is Wangda
Zongga: centre of Gyirong Xian